Jochen Bockemühl

Kari Järvinen

Extraordinary Plant Qualities for Biodynamics

Extraordinary Plant Qualities for Biodynamics

Jochen Bockemühl

Kari Järvinen

Translated by David Heaf

Floris Books

Translated by David Heaf

Originally published in German as
Auf den Spuren der biologisch-dynamischen Präperatepflanzen
in 2005 by Verlag am Goetheanum, Dornach, Switzerland
This edition published in 2006 by Floris Books

British Library CIP Data available

ISBN 0-86315-576-6
ISBN 978-086315-576-5

Produced in Poland by Polskabook

Contents

6

Contents of text boxes

Introduction

A different involvement with the living world

We sometimes notice how a friend's approach to life is reflected in their home surroundings, for example, in the kind of furniture and pictures they have and how they arrange them. In a similar way we can know the basic attitude of people living on a farm from what picture is created by the entrance, or the fields and their surroundings, or by how the manure is managed etc. The effect of that attitude goes as far as the yield and quality of the farm produce. On further reflection we realize that the life of the whole earth is ultimately guided by the attitudes of human beings and the actions resulting from them. But to what extent are we aware of the effect of the picture that we create?

When we make something with a particular aim in view we are naturally awake to things and processes and sense that our deeds are embedded in the reality of material facts. But, in the circumstances of the living world and human life in which we live, the consequences of our thinking and doing largely escape our notice, which is concerned more with external things and success. Many now regard this as the only acceptable stance for people living in a world of objective facts. We try to cultivate the onlooker's attitude of mind and from this vantage point deal with the 'unforeseen side effects' of our actions. When a trade journal urges a 'new performance drive' and offers training with games to motivate a company workforce it stems from the same attitude. It completely ignores the personal lives of people and even the nature of what is produced. It merely comes down to productivity that can be expressed and calculated in numbers.

In the days when a familiar landscape, the social life of the community and religious attitudes once provided the framework in which life unfolded, and each person's disposition was a part of the whole, it was still natural for people to assume that others thought, perceived and acted in the same way. But today all that has changed to the extent that both perception and feeling have lost their vividly-experienced natural and social contexts; thus, their significance for knowledge has also diminished; scientific investigation has replaced them. It has to assume that what is to be investigated is a self-evident certainty. Thus people trust the authority of a blind study by indifferent observers more than their own percepts and understand them by *thinking*. A phenomenon that accompanies this is the longing for pictures that stimulate our emotional life. The media takes care to bombard us from all sides with pictures taken out of context. Without our awareness, these pictures interfere with our natural relationship with the world. Their effects continue into the attitude that governs our life and thus into our deeds. We are, of course, well aware of how this is used in advertisements and fashion trends. Pictures with the greatest impact are used to affect us deeply by association and replace the pictures of former times that created connections with real life. Only self-reflection can tell us to what extent this is influencing our own lives and unconsciously shaping our way of living and thus our environment. Without such reflection we will certainly change the world with every thought and deed, as we see fit, but we will have no real involvement in what we are doing. Self-reflection here refers to our own relationship to how much and in what way we relate our intentions to a world that we are usually accustomed to regarding as external. How can we now achieve this new involvement?

During Whitsuntide 1924, in a lecture course at Koberwitz, Rudolf Steiner sketched a new concept of agriculture. In it, he drew attention to the way in which people could maintain a living connection with nature through their farming activities. He expressed this in pictures that spoke to the feelings and which for decades have informed the practice of farmers engaged in biodynamic agriculture. Since that lecture course, consciousness has moved further in the direction already discussed in a way that at the time would have been barely imaginable. This makes self-reflection all the more urgent. The Agriculture Course contains many elements that show how an attitude of self-awareness can arise, but that also requires a new understanding of what is presented.

(Geisteswissenschaftliche Grundlagen zum Gedeihen der Landwirtschaft (GA 327), *Dornach, 7th edition, 1984; trans:* Agriculture Course: The Brth of the Biodynamic Method *(London: Rudolf Steiner Press, 2004)*

About this book

New approaches to understanding Rudolf Steiner's Agriculture Course have emerged over many years of engagement with it, from meeting farmers working biodynamically in different landscapes around the world, to holding regular courses with people involved in agriculture. With this as a basis, we would like this book to make people aware of a necessary extension to the current attitude to life that will enable them to co-operate practically towards the healthy evolution of the earth.

The use of preparations derived from healing plants plays a key role in biodynamic agriculture. In this book the plants are used as an example to illustrate not only how a new and practical relationship with the preparations can be forged, but also how people can gain a new and authentic relationship to nature.

As a result of this, the *effect* of the preparations, something that is not easily accessible to the modern mind, will acquire a totally different reality. What in the world of technology we were accustomed to regard as the 'side effects' of an intended procedure, will, from this new way of looking at the subject, become the desired effect which results from the connection between the life of the earth and the human being. However, agriculture can also point the way in other areas of life, if we reflect on the special method of agricultural production involving its mysterious 'means of production' — field, meadow, cow — and learn to understand them as organs of an organism.

We, the authors of this book, are of course on our own path, with our own insights and attitudes. What we are aiming at here is to encourage readers to view their personal relationship to the world with greater self-awareness. This involves first and foremost learning to look in a new way at the natural phenomena we have to deal with and rethinking our own relationship to them. This can then become the basis for working with the preparations independently.

We begin the study of them with simple observation exercises that enable us to experience how we ourselves are deeply connected with outer reality. Then we move on to considering landscape in order to become aware of how differently things speak to us when we grasp them as a whole compared with when we regard them just as objects or substances for our own use.

The next step is to clarify the landscape contexts in which the preparation plants grow. Through using what we have learnt, we then try to find the character of the plant from the way it manifests itself in relation to the landscape. Finally, against this background, both the seasons as a whole and the formative processes of the preparation plants become gestures that speak to our experience.

From this we try to derive an insight into the healing effects of the preparation plants on human organs. This calls for a deeper understanding of what organisms and organs actually are and how the realms of nature live differently with the four elements of antiquity. The essential thing in this approach is not to have ready-made opinions, but to experience the *way* to this understanding through which, in dealing with the preparations, we increasingly familiarize ourselves with the contexts of life on the earth.

In the case of some of the concepts, we use text boxes to explain the direction in which we would like to guide the reader's attention.

At this point we should like to express our warm thanks to all those who have contributed to the creation of this book.

The picture of the human being as a microcosm

The traditional wisdom that the human being is a microcosm in which the macrocosm is harmoniously integrated is no longer taken for granted. Such a picture is less and less part of human consciousness.

Yet it could be a model for an aim in human life to seek out the cosmic connections of world phenomena and, though our own endeavours, actively immerse ourselves in them. It could generate a new way of thinking in the midst of the dying processes of the earth and in the face of all ideas valid for consciousness of objects and of what is calculable. People could seek new personal involvement and fully conscious participation in the course of evolution of the earth. They need to devote themselves lovingly to world phenomena and, from this, form the earth of the future. Such a formation process cannot happen from outside. As with all things in the organic realm, it must grow from within. Conscious participation means letting the germ of a new life of the earth arise in the human being.

1. The World of the Senses can Become Transparent to Experiences in Soul and Spirit

Let us start with the first simple exercise that can lead us to an understanding of what it means to become involved in natural phenomena. For this it is first of all necessary to *contemplate* how what lives in thoughts, feelings and will impulses is connected with the world we perceive in such a way that the reality in which we exist results only from bringing both sides together. Then it is a matter of experiencing it ourselves and living with this experience. The following exercise with a branch and the resulting experiences can be understood by reading the text, but the experiences can only be obtained oneself if the exercise is carried out with a real branch that someone might have given us. It is a worthwhile experiment as it forms a basis for all the following steps on the way to insight into the effects obtained when dealing with the preparations.

The picture of the phenomenon of a branch in winter

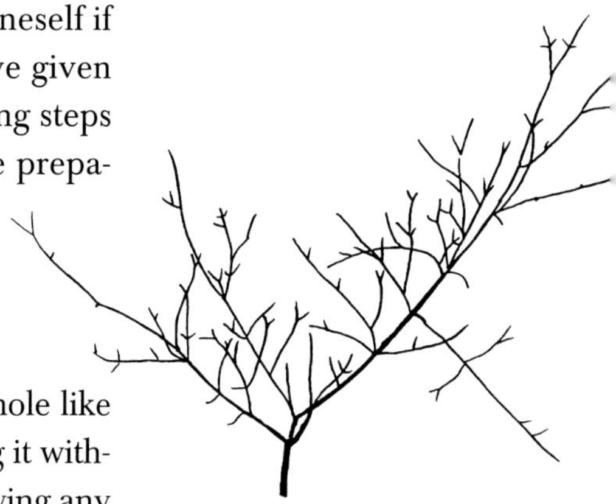

We take a bare branch severed from its original context, study it as a whole like a picture or a sculpture and try to devote ourselves actively to perceiving it without remembering or interpreting or going into details, even without drawing any conclusions from the initial mood that it provokes in us.

Figure 1

If we rotate and tilt the branch we will discover, by means of our aesthetic sense, orientations that are more satisfying or less so. After a while, with a little patience, we will arrive at just one position that we find completely satisfactory. Anyone who actually tries this exercise will be able to experience this.

Pictorial (imaginative) perception

When we look at a picture that has been painted, we can of course ask what physical and chemical forces have participated externally in its creation. But in this way we shall never discover what the picture is expressing within itself. A work of art as an object (composed of separate objects or illustrating an object) only really becomes a picture when we focus on what it points to as a whole. A painting is also a single object in space. By focusing our imagination on the pure phenomenon we can become aware of the whole — the spiritual source — out of which this single object has arisen. In that way we might experience something of what was in the mind of the painter as he painted the picture.

A natural phenomenon as a picture — that is, grasped imaginatively — leads us in a certain way to an aspect of the cosmos or to the language of the inner nature of what we are looking at.

Thus, purely with our aesthetic sense we can find, in what the branch looks like, a special orientation without knowing anything about the natural surroundings from which the branch was taken. Yet we find this now corresponds exactly to the original orientation of the branch on the bush from it came. When we go back to the bush it is easy to find the right position in which to replace the branch and thus complete the overall shape.

The whole process gives us an inner certainty. We understand the way in which we grasp with our artistic sense, purely through experience, what produces a unique orientation in space from a form that has arisen out of life.

Identification through understanding

An experience of how, in the process of knowing, we identify with the things of the world emerges from the seemingly simple question of how we recognize a thing. For example, what form does the shape of a cube take in our mind before we look at a crystal of cooking salt?

What lives in our thinking is neither perceptible to the senses nor can we picture it in our mind's eye. Pure geometric thought activity is an organ of perception for spiritual relationships. This enables us to picture cubes at will independently of their size, mass or orientation. If we see a salt cube our thinking is the organ we use to contemplate it from all sides simultaneously. We complement the percept with thinking and give it its meaningful context from the perspective of geometric form. What we recognize in the shape of the crystal goes beyond physical perception, otherwise we would not be able to tell that one corner is missing or a surface is deformed. Here, with mobile thoughts, we are dealing with the possibilities of cosmic order, for which we develop a sense in pure, geometric thinking. However, it will not help us come close to the taste of salt or its other properties.

How did we arrive at this first experience?

Obviously we identify ourselves so strongly with the appearance of the branch that we see in it, and through it the harmonious relationship between a sphere experienced pictorially and its vertical growth that we experience in our own bodies as well. With it we immerse ourselves in the outer spatial orientation of the branch. Only as a result of this does the possibility arise of completing the overall shape of the bush.

Thus we have learnt the first basic principle of all life on earth: *that we do not meet any plant, animal or human being without at least grasping pictorially the overall impression of how it reveals something of the particular way in which it exists. This happens through our relating what it looks like to our own physical experience.* With sufficient attention, we can make ourselves aware of this in every life situation. In so doing we will notice something of the inner relationship of the realms of nature to the cosmos and to the earth's spatial environment. An essential feature of the realms of nature is that these relationships are different in each case.

Becoming aware of the traces of life in the branches

Besides the original orientation of the branch other phenomena can be studied using additional branches that point to other dimensions of reality (see Figures 1–8).

We notice stiff upright growths and various bends. We experience the forms as movements that we trace inwardly: in Figure 2, for example, an upwardly reaching curve; a gentle bend actually enhances the impression of its smoothness; in Figures 3 and 4 indefinite, weak movements, often dropping downwards. With Figures 5 and 6 the course of the twigs is more hesitant, only in different ways: in Figure 5, densely compressed, in short sections; in Figure 6 more meandering.

The movements are interrupted rhythmically in varying ways by nodes. Thus Figure 2 shows very long sections, often of more than ten centimetres,

Figure 2

For cosmic perspectives, see page 53.

Figure 3

Figure 4

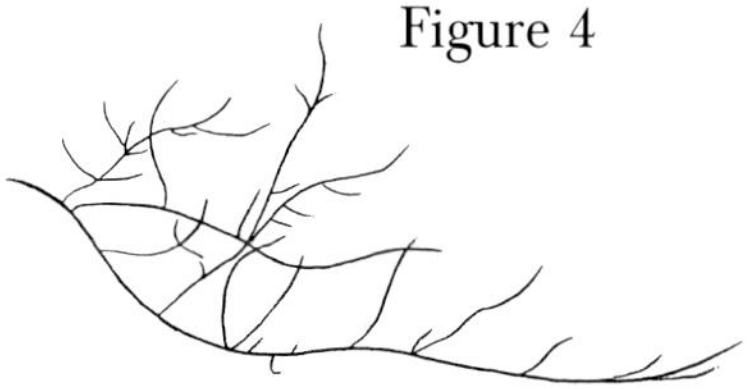

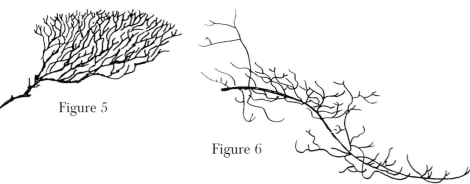

Figure 5

Figure 6

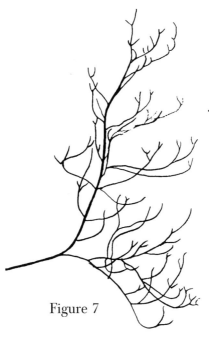

Figure 7

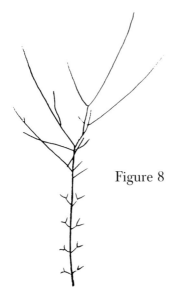

Figure 8

whereas those of Figures 5 and 6 are very short. In addition, Figure 1 shines distinctly red on the upper sides and yellowish green underneath; in Figure 2 the whole thing is red, but Figure 3 is all yellowish green. Figure 5 has red tips and Figure 6 hardly any reddening.

The rough, blackish-brown composition of the thicker part of the branches, especially in Figure 5, gives us the impression of their becoming rigid and brittle in contrast to the smoothness of the thinner twigs on the same branch or to the shoots in Figure 2. The individual properties say something not merely about themselves, but rather point to a unity like a stream in which each of them flow together.

If these observations prompt us to speak of growth, age and youth etc., then we are identifying with neither an object nor a picture but with a *process.*

Impressions of phenomena grasped in transformation come together in our encounter with the branch. With this we are developing an elementary capacity for contemplating *life processes*. This method of experiencing can seem like looking at a flowing brook. Just as the many moving reflections and transparencies convince us that the water is flowing, so too, through the inwardly flowing relationship of the branch qualities, an experience unfolds a living process that has left its traces in it. This plant-like appearance is different from that of rigid, mineral things which we picture as almost timeless, like a crystal with particular permanent properties.

Formative forces; perceptive thinking

If we here refer to formative forces in connection with the etheric we need to note carefully to what experiences the word 'forces' is attributable.

The 'spiritual picture' from which a painter works is an intention, a guiding idea, a perception that lives in the totality of experience but cannot be visualized. Ideas exist in a world out of which a variety of representations can be developed from a particular spiritual perspective. A finished painting is a unique work of art. It is witness to an actual interaction between the intention and the material in which it appears. Painting ultimately happens through physical-mechanical forces brought about by the limbs, but these are guided by the mind of the painter so that the intention, the idea that is living within him, shines through the work of art as a complete phenomenon. Thus the physical forces are subject to the painter's efforts that as a whole involve soul-spiritual activity.

In the living world such an (intentional) formative principle acts outwards from within, and we participate in it with our perceptive thinking. Thus, formative forces do not work separately, like mechanical forces of the limbs or body, but holistically; that is, in the way we have them and use them in perceptive thinking. It might be better to use another word for these forces. But when by force in general we mean the will element, in the same way we can find it in thinking and there deliberately distinguish the mechanical, detail-oriented thinking from the holistic kind that we adopt in contemplating something aesthetically. Here, we have called the latter *perceptive thinking*.

Experiencing formative forces

In following the traces in the branch we have perceived *transformations*. Through the inner activity of doing so we have entered into a movement that requires our active participation. It allows us inwardly to participate in forces from which the forms of plants are produced. These forces make connections in the same way that perceptive thinking itself is active in making connections. It is only the results of these activities that are visible in the pictures of the sense world.

In order to find an orientation in space, we are dependent upon our own physical experience and the relationship it has in space to the earth and the surroundings. Real formative forces can be experienced by participating with our own life processes, as we know from seeing a yellow *after-image* emerge after looking at a blue surface. Thus we recognize a second elementary principle:

For after-images see page 16.

As well as what physics recognizes as forces — that is, those that we make use of physically in space with our limbs in lifting, carrying and pushing — there is the picture of a whole appearing to the senses. This discloses itself to our aesthetic, artistic sense. In perceiving and experiencing, we can recognize forces that allow the substances of organisms to arise and take shape coherently from within.

After-images

The percept of a blue surface only becomes reality for us when a particular sensation is congruent in intuition with the concept blue. This concept subjectively lifts this particular kind of sensation out of a general experience of colour. Each person has to do this for themselves. Our whole life-organization participates in this; that is, the organization that makes up our living body, that holds it together both in itself and in relation to the world. With a little care this is easy to observe. If we concentrate for a short time on the pure impression of blue and then look at a neutral surface, a yellow after-image appears. This is a specific sensation that we can agree about just as much as we can with the blue.

After-images, that have been illustrated here in the simplest form, teach us how our organization responds to the percept of blue with a yellow that lights up from within. The single phenomenon of a colour thus acquires in us a position in the overall context of colour phenomena that is just as objective as summer is in the seasons of the year.

Effects of the earth and its surroundings

Many of the observations presented here were already taken for granted at the beginning of the exercises when we looked for the correct orientation of a branch. Otherwise it would have made no sense to speak of upwards and downwards movements. Thus there is a definite relationship to the earth for all the qualities of life processes that are expressed in the picture of the phenomenon. But this is much more comprehensive than in the first step of the exercises. It includes feelings we have in our own physical life; those of standing upright and bending down in gravity while in the freshness of youth or when becoming old; of being heavy and light; of being firm and soft and of being massive and sparse.

For aspects of the cosmos, see page 53.

In addition to the relationship to the earth there is also a relationship to the sky, one that can make a plant appear more or less impressive. When we use terms like 'light' or 'bright' or 'shady' when looking at a plant we are referring not just to lighting conditions at that moment but also to what its atmosphere has contributed to the way the plant is formed and through which the plant appears as a picture of the forces that create the relevant conditions for it. This has gone into its substance and gives rise to its quality. Here, our bodily experience of an object is permeated by a cosmic wholeness; with something we live inwardly and experience only with perceptive thinking.

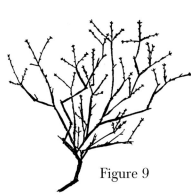

Figure 9

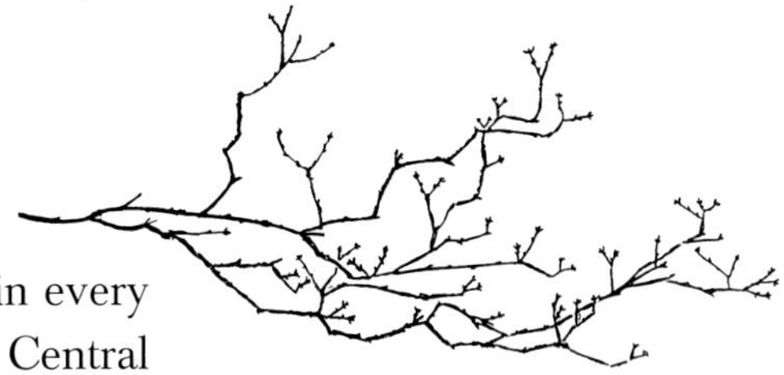

Figure 10

Tree species and their language of gesture

Although each individual branch conveyed something very different, in every one of Figures 1 to 8 we can recognize dogwood, a species native to Central Europe. By studying many branches we develop an increasingly comprehensive capacity to perceive the unity of character that speaks through all individual phenomena and life processes. The character of dogwood is fundamentally different from the character of a species that tells quite another story, for example the angular, nodular, partly brownish-silvery gleaming branches of oak

(Figures 9 and 10). Experiences awakened in this way affect us more deeply than those of the life processes.

Immersing ourselves in the phenomenon at this level leads us to gestures that express themselves to our perception in the particular spiritual direction from which the form emerges. We become aware of what the visible plant is actively producing. Each expression of a plant is an effect of what was in the general experience of life processes, only now it can be grasped more concretely as the character of the species.

This brings us to a third elementary principle: *each concrete form of a plant is not only the effect of life processes, but is also permeated simultaneously by the activity of a unity that, like a tree species, speaks to us through its manifold appearances from a particular spiritual direction.*

What we have learnt in this chapter can be summarized as follows:

First of all we learnt, through experiences in our own body and inner capacity for perception of the appearance of a branch, how any branch can be uniquely incorporated into a spatial relationship between the earth and its surroundings.

Careful observation of the growth and division of the shoot led to an experience of formative life processes that showed us their effects pictorially in form and colour. They became imaginative experiences.

By searching for the unifying 'character' in various samples of a plant species, we encountered a new spiritual disposition of soul. We experienced *intuitively* what the life processes had formed *in precisely the way* they appeared in the individual sample.

The exercises are especially demanding when bare branches are used. Using the traces in the winter branches, the really important thing is feeling our way towards creating a vividly coherent inner perception of the growth of the shoot. Out of that, an organ develops for perceiving the formative effects in the overall life of the branch or bush. It makes us look forward in the following year to observing more closely how these effects manifest themselves in the production of leaves, flowers and fruit, as well as in wilting.

A special feature of woody plants is that their trunks and branches leave behind traces of the life processes into whose forms and colours everything has been poured and through which can be experienced what previously has appeared in the growing season and influenced them. Herbaceous plants leave behind no winter traces of that kind, but the life of each plant somewhat imprints itself on the earth in a more hidden way, for example in the transformation of the rhizome and the soil conditions. Winter traces appear more obviously in the process of succession in the formation of plant communities.

2. Awakening a Sense for Landscape Moods

When we perceive plant form we are connected with not only the sense world but also, if we take note of the specific circumstances, the atmospheres in which the form arises. In this chapter we draw attention to how we can become aware of these circumstances, that is, how what we see is dependent upon how we look at it and how the external appearance of the landscape is connected with what stirs within us, as well as with plant form.

Landscape moods; atmospheres

With the mood of a landscape or some other phenomenon we mean not only a feeling that arises in us but also its unity with the sensorial-aesthetic percept. The exercises in the last chapter indicate how a feeling can become an accurate organ of perception for qualities, if the relationship to the sense percept is kept in view with perceptive thinking. We do not stand back from it, but stay close to it — that is, we ourselves are active in combining outer and inner in that we switch our attention back and forth. This method of perception can be described as atmospheric, because the atmosphere is the mediator of the soul aspect at work in the world of the senses, or of the cosmic-astrality with which we unite ourselves while perceiving and experiencing.

On a walk in the countryside during early summer, we receive many impressions one after another. We have barely finished admiring a colourful meadow in flower, when we come to woods which at first seem dark. But in the woods we soon adapt to a very different way of perceiving, no longer seeing a complete and varied coherence. Instead of experiencing the flowering meadow — which of course forms the background for observing butterflies and flowers — our seeing and experiencing enters the darker, green-brown surroundings into which we have plunged. With each new perceptual situation, our interest in individual things awakens differently in accordance with the overall experience.

Usually we hardly notice these changes in mood in successive places, and they are, of course, not *our* moods. The plants manifest themselves in a completely different way. They have grown in the atmosphere of the place and their form belongs to the expression of that particular atmosphere.

Meadow and wood habitats are so different from each other that each frequently rules out the occurrence of the plant species of the other. Wood and meadow species can meet at the transition points. But there are also species such as herb bennet (*Geum urbanum*), ground ivy (*Glechoma hederacea*), stinging nettle (*Urtica dioica*) and valerian (*Valeriana officinalis*), that wholly express the character of these transition zones in the picture of the phenomenon.

Our diminished attention for the atmospheric context of a phenomenon has the result that we recognize the same plant species at different sites far more easily than we do its differentiation in growth. But when we pick a sample from one site and put it beside one from another site the differences suddenly become obvious. In each case we did not notice them previously because at the original site all phenomena harmonized to the extent that the species stood out more clearly than the variant forms in which it appeared.

Differences in colours, forms and compositions of a site are very diverse. If we pick out details as examples, compare them against the background of the whole and raise their relationships into consciousness, as with experiencing a picture, then there arises a sense of the atmosphere that helps the plant to grow into the particular appearance under consideration. In addition to this, it becomes clearer how something atmospheric is given out by the picture of the phenomenon of a plant that is united with the atmosphere of a particular location. Awareness of these kinds of relationships leads to an insight into the intimate natural interactions that are referred to in the seventh lecture of Rudolf Steiner's Agriculture Course and on which a new approach to agriculture depends. An inner complement is added to outward appearance, which only then becomes a whole. This 'greater intimacy' contrasts with the 'broader interactions' that alone are often the sole consideration of organic agriculture which reckons with the quantitative relationships between cattle, fertilizer, fodder etc. in relation to yield, and which thinks in terms of cycles of matter.

For stinging nettle in light and shade, see page 22f.

1   2

3   4

Legend to the adjacent illustrations. Stinging nettle in light and shade

Two examples of stinging nettle (Figures 1 and 2) together with the environments in which the two plants have grown: Figure 3 sunny pasture where the nettles reached well above the grasses; Figure 4 semi-shaded site in a wood.

How can an aesthetic judgement become relevant to ecology?

An individual plant can be looked at like a work of art. For example, we see and experience the appearance of stinging nettle in Figure 2 with its seemingly floating, broadly outstretched, delicately green leaves. But we also see the beauty of another, upright nettle (Figure 1) with its stiff, red stem, delicately pink florets and dark-green leaves folded at the tip. Each nettle has its own character. This is increasingly richly experienced and communicable to others through paying attention to detail in relation to the whole picture, for instance the different gleam of the stinging hairs, the density of leaf veins, etc.

In this way we are additionally enriched with the wealth of special experiences with a plant that perhaps we do not normally pay attention to. We can now develop a preference for one or another nettle. By thus making a closer and closer link between our experience and the stinging nettle, through the resulting intimacy we can achieve an awareness of our own inner orientation to the world. In this way, just as with a work of art, studying a plant contributes to the cultivation of the inner human being.

A work of art studied for its own sake leads to particular experiences, on the basis of which the human soul is formed. We could put the work of art in the garden and notice how it harmonizes with the surroundings, or arrange it in accordance with them. In doing so the work of art is not yet relevant to the natural life of a place we are landscaping. For that to happen, it is necessary to extend the study like the one here with the example of the plant and in the picture of it as a phenomenon, arrive — as we did with the exercise with blue — at a kind of after-image with which we inwardly experience and perceive light and shade, moistness and dryness, fertility of the soil, etc. This after-image formed according to the phenomenon of the plant can be so precise that it coincides with the actual surroundings of the plant.

Pictures of landscapes

The following pictures are intended to help us attune to the experience of the atmosphere of a place. Initially, we receive an overall impression to which we quickly react with sympathy or antipathy. Individual objects are more or less clearly visible. In order to grasp more clearly what is expressed by a picture as a whole, it is advisable to refrain from labelling any objects in it with concepts, or react emotionally by liking it or not liking it. What we are looking at can only really become a picture through this restrained but alert, open kind of looking.

In artistic contemplation there is a method for properly clarifying the sensation of an atmosphere that is difficult to grasp at first. It involves scanning individual areas of a picture in such a way that they once again unite in perceptive thinking as conscious, pictorial relationships. While we remain engaged with the phenomena, thinking in this manner and experiencing our own activity, in our soul arises a richer perceptual impression of the atmosphere. The activity of our own consciousness puts itself at the disposal of contemplating the pictorial relationships of the percepts so that space is made in the soul for the inner nature, the being of the object of study.

Thus the being is revealed in various relationships: an object to light, that is, the illuminated environment; a plant to the surroundings that produced it; an animal to a world that it seeks in its environment, and in doing so shows what is living in its soul; our own way of thinking and doing and its consequences for landscape.

Insight into the being, towards which we turn with the deepest of interest, grows in proportion to our attention to our own activity that awakens within through the manifestation of the being. It begins to speak to us through our own inner orientation to it.

A wildflower meadow

When we approach a place like the one illustrated in Plate 1, while attending to the atmospheric aspect as described, impressions arise in us such as fresh, light, airily mobile lightness, cool, but also in contrast dappled shady sun.

The view awakens memories of previous experiences: warmth, approaching rain, perhaps birdsong, the humming and fluttering of insects resonating in the consciousness. They awaken the mood of a sunny day in early summer.

We know how such a picture of the phenomenon has come about. It is an expression of people's particular activities: forest plantation, mowing, or grazing cattle in a particular rhythm in harmony with the seasons. The first mowing of this meadow has not yet taken place.

The surrounding open, broad-leaved woodland hinted at in the picture as

Plate 1

well as the dense, lower spruce forest also belong to the context of life in which this meadow is situated. The farmer is taking care over the quality of his forage.

A note on methodology

With this way of picking things out of a picture and letting them come alive in us, there are of course obstacles to repeatedly looking without prejudice and discovering something new. These have to be overcome. For instance, it is known that the first settlers in Australia actually saw European landscapes there and painted them as such. Furthermore, by developing the land according to the atmosphere of their place of origin, for example with golf courses, and introducing plants, they established around them an environment that is alien to the landscape of Australia. They only imported the things they were used to, which continued to have an influence and produced unwanted new effects. Only gradually did the settlers discover what was special about their new landscape. Aborigines, who learnt painting from the settlers, found it far easier to capture the mood elements of *their* landscape and properly illustrate them. We learn from this that prior experiences indeed produce immediate links between experience and a phenomenon, and that through them we are prepared to see landscape as a varied unity. The gaze prepared in this way is guided to many things that we would not otherwise see. But as a result, in the face of experiences we bring with us, we need to open our eyes to the different quality of new impressions, and to enter into the special moods and even into the spirit of a place that we are visiting.

Furthermore, our haste to conceptualize and our susceptibility to unexamined emotional impressions are also obstacles. Both of these are mental associations which happen in cognition in an encounter with a phenomenon in a way that is similar to when we lose our footing because of momentary inattention. It is therefore a matter of consciously slowing down the journey to perceptive experience of the spirit in the phenomenon so that we can pause awhile. We can vividly reveal the inner nature or being of the object of contemplation by constantly taking care to keep awake our relationship to the physical phenomenon with our organs of perception in soul and spirit, and by making judgements but also withdrawing them again. This slowing down leads to an enrichment and clarification of basic experience from within. Through such differentiation of our involvement with phenomena we regain what gives us inner security. This increasingly replaces the 'home' we have lost. We feel at home in a world which has our interest.

Contemplating the relationship of individual phenomena in looking at the whole

After these remarks we return to the picture and turn our attention to the colours and how they are arranged. The light blue at the top contrasts with a light green interspersed with yellow containing various configurations and shapes of red, white, yellow and blue spots. A thinner band of dark-green, open, vertical stripes undulates across the centre of the picture. Above that float some rain clouds stretching from the horizon into the blue, and under the dark-green a delicately reddish haze hangs over the light green below it. To right and left are various dark-green tones that connect the upper part of the picture with the lower, framing the whole.

Now we can include the things that drew our attention to the details. What place do the individual plants have in the whole picture, the bright-red zigzag clover (*Trifolium medium*), the bellflowers (*Campanulae*) and the rose bushes? What do they add to it and what conditions do they need to appear? Could the red clover play a part in the colour composition without the woodland margin?

The more precisely we examine the details and, as a result, include them contemplatively in the overall impression, the more certain we can become that this is not a picture of a habitat in southern Germany. Our impression of it instead leads us to a world of long summer days in central Scandinavia. The joy of the Swedish midsummer festival floats in the air, so to speak. Even if we have not been there, we can sense it.

Landscape affects the human soul. We can develop an increasingly differentiated receptivity to what already resounds in the picture of a landscape. This way of recognizing subtler effects in a landscape is fundamentally different from talk of the 'energies' experienced in them after a fleeting glimpse.

Rock vegetation

The picture of vegetation on rocky ground appears less green (Plate 2). Delicate plants form a kind of natural rock garden. In between them we can see rocks all around. The landscape is mountainous, stony, with sparse, scattered vegetation. Herbaceous plants grow very close to the ground. There is little recognizable leafiness. It is all very colourful: the brownish-red and yellow flower of stonecrop (*Sedum*); the green is partly bluish and partly turquoise or dark brown-reddish. The flowers and the leaves below fan out in a pattern. In fact there are hardly any herbaceous plants to be seen.

Only the scrub is dark green. It is gnarled and thorny with leathery leaves in places. On the bush in the right foreground, not only are the flowers red but the stems are too. Overall, the bushes together with the cypresses in the background point us to the south, for example to limestone mountainous steppe vegetation in the foothills of the southern Alps.

All these comments on the picture are based on impressions which can only be expressed directly by way of hints: the predominant impression is hotness and dryness. To this is added the experiences of brightness, barrenness. These hints cannot be made more specific, but they help us to attune our perception to the individual relationships of the gnarled scrub to the limestone rocks; how and where the red colours appear at characteristic positions. Thus gradually this can lead a person towards experiencing a good, though seemingly external, description of the phenomena and their relationship to the whole. With this approach we are beginning to experience the being of the place or the *genius loci*, as it was once called.

This being we experience comprises the soul-spiritual surroundings into which we have entered and which, like a human being, cannot be encountered concretely. As we are not accustomed to relying on this sort of experience, we are inclined to force it into a scientific conceptual framework and thus end up in abstractions.

Plate 2

Landscape biographies

How did the meadow illustrated in Plate 1 (page 25) come about? It is not natural. The picture it now presents as a phenomenon contains a biography. The forest, that we would expect in comparable situations and which would be regenerating itself, is managed; the cattle have grazed the pasture and occasionally it is mown etc.

In contrast, Plate 2 on page 29 allows us to experience a very different landscape biography. In earlier times it was part of a neighbouring olive grove that was later chopped down, and after over-grazing and increasing soil erosion was finally left fallow. The wood is not returning here. At best maquis brushwood takes its place. The thin but extremely fertile black humus in the cracks in the rocks provides the context for the richly flowering, radiant vegetation. The arborvitae cypress trees (*Platycladus orientalis*) in the background also make a significant contribution to the overall mood.

Plant and animal aspects

The spontaneous occurrence of a plant species shows us specific qualities in the context of change of a place. Details of the plant, such as coloration, scent and taste, appear in connection with the atmosphere of its habitat. A growing, flowering plant develops in a breathing relationship with its concrete surroundings. It achieves a climax with flowering and then fades again. Conversely, however, the moods of a landscape in which particular plants occur enable us to experience special characteristics of these plants.

Landscapes are pictures that affect the soul of man and animal. The woodland surrounding the meadow in Plate 1 is an important factor for insect life and contributes to bringing about a healthy diversity to it.

The second landscape, largely relinquished from the care of the farmer (Plate 2), is associated with an animal life quite different from that of the neighbouring olive grove (Plate 3), and has an enriching effect on the latter.

Plate 3

One aim of such contemplation is to sharpen our vision for landscape qualities and then in a particular place, for example on a farm, to bring about a diversity of specific qualities that can send out healthy interactions at neighbouring locations. From what is provided naturally where wild plants are managed, a diverse wholeness should arise in such a way that cultivated plants and farm animals can thrive with the support of their surroundings. An essential part of the whole farm environment is the attitude of the people in charge who are concerned with these matters.

On our way to understanding how the wholeness of a place is associated with a particular plant species and what this place can produce, we shall try in what follows to grasp with pictorial experience the atmosphere of the natural habitats of each plant that Rudolf Steiner indicated for the production of the preparations for treating compost and the ground.

It involves widely distributed wild plants that are also known for their healing properties: yarrow (*Achillea millefolium*), German chamomile (*Matricaria recutita*), stinging nettle (*Urtica dioica*), oak (*Quercus*), dandelion (*Taraxacum officinale*), common valerian (*Valeriana officinalis*) and common horsetail (*Equisetum arvense*). Each of these plants has its own specific environmental conditions, which we shall try to clarify in the next chapter.

3. The Habitats of the Preparation Plants

The special spatial and temporal qualities of the environments that allow the preparation plants to thrive in the landscape enable them to develop specific substances in the context of life on the earth. In this respect, with the components of the plant preparations we take cosmic qualities from the earth, qualities that are closely connected to what we experience in the atmosphere of the site at which they grow during the course of the year. We will try to wake up to the special characteristics and practical significance of these qualities in order to understand how they can help us intensify the individualization process of agriculture at a particular place.

With the exception of oak and common horsetail, the preparation plants are harvested in the season in which they flower. This gives harvesting a natural sequence and takes us to different places. In the illustrations that follow, in each case on a sunny day that stimulates flowering, we should like to encourage readers to look out for the qualities of the atmosphere.

Dandelion, April to Early May, fertile meadow with tall oatgrass

A wealth of shining yellow flowers amid the deep green of an open meadow landscape evokes *joie de vivre*. In between, a bluish, whitish haze of lady's smock (cuckoo flower, *Cardamine pratensis*) brings a cool touch to Plate 4. Above them, light and airy cherry flowers open up and brighten the scene, but this contrasts with the black branches of the trees. The fresh green gives the impression that the grass is very appetizing.

The earth sends out an intense agility. When the sun goes behind a dark cloud in the evening, the yellow of the flowers radiating from the ground becomes obscured, but re-emerges the next day with the sun.

The impression of agility in a flourishing dandelion meadow is enhanced by something else. If we rest our eyes on the shining yellow amongst the green and then move them a little, some dark patches of blue-violet appear fleetingly in between. At most we usually notice the slightly mobile impression of the dandelion flowers as earthly images of the sun. Because of the intensity of their colour they too, like the sun itself, have after-images that leave their mark on that overall impression.

Our eye for the special features of a dandelion's habitat is enhanced by turning our attention to the transformation of the fertile meadow as illustrated here. This later flowers colourfully, and turns into a luxuriantly growing meadow, for example, where, after the lush flowering of dandelion, the most noticeable colours are the different yellow of meadow buttercup (*Ranunculus acris*) and the more open whiteness of hogweed (*Heracleum sphondylium*) and cow parsley (*Anthriscus sylvestris*). When dandelion grows with plantain (*Plantago*) rosettes on paths that cross less fertile ground or on dry sites, it remains small and the green often appears tinged with red.

Plate 4

Stinging nettle, end of May to June, damp woodland margin in semi-shade with a transition to water meadow woodland

It is usually hard to tell when nettle is starting to flower, as its flowers are so inconspicuous.

Green on green appears in this illustration of the meadow woodland (Plate 5). It is darker at the top, more or less depending on the leaves of the trees. At the bottom it is structured by the uniform enveloping foliage of the nettles. This structure extends into the more shadowy parts; in the brighter parts it is denser and more prominent. There the stems stand out separately and more distinctly with their sometimes reddish tones. The grey-green of the flowers increases in the upper region of the stems.

The water meadow mood is at its most intense in semi-shade on river banks and the edges of streams that are frequently inundated by spring floods, and where organic material piles up. It is also noticeable how small clumps of nettles appear near compost heaps and on the margins of woods and fields.

Plate 5

German chamomile, beginning of June to July, edge of a grain field

Chamomile begins to flower on the margins of fields and lanes when the meadows are fully grown and the grasses and wildflowers are flowering in increasing diversity, for instance, similar to the flowering meadow shown in Plate 1 (page 25).

In Plate 6 (opposite), the situation is one of warmth and brightness. The grain is already ripening. A mood reminiscent of the Mediterranean region allows colourfully flowering plants that originally came from warmer countries to thrive, such as chamomile, poppy (*Papaver rhoeas*) and cornflower (*Centaurea cyanus*).

In farmland, chamomile is found amongst cereal crops, between root crops, in gardens, on the margins of fields or farm tracks, as well as on rough or derelict ground. In order to flower, chamomile needs places where the ground remains open and has a tendency to crust over.

Thus, chamomile is in its element in the Puszta of Hungary. There it covers vast areas of the extensive pastureland which has a tendency to hold puddles after heavy rain, and the soil becomes saline and encrusted to the extent that no grass can grow. It is here that the small, openly-branched and seemingly dif-fuse, richly flowering chamomile bushes shine in the June sun. Their green is light and delicate and amongst it numerous flowers float like yellow-white stars.

Plate 6

Common valerian, June/July, woodland margin undergrowth with perennials, changing to tall grass

The flowers of common or officinal (medically used) valerian appear at the same time or shortly after chamomile, but their flower umbels only develop gradually to full size. Harvesting is best done when the sweetish-dull flower scent is wafting from them.

In Plate 7, where the woodland margin projects into the meadow, the mood is somewhat lighter than a typical nettle habitat, yet is nevertheless rather shady-moist and frequently somewhat musty.

A valerian atmosphere can develop on the open margins of woods or hedgerows, in clearings or on river banks bordered by trees or bushes — that is, mainly where shade makes a transition to sunshine, or moistness to dryness. These places tend to become overgrown with scrub and tall herbaceous plants. In spring they are light but they close over towards summer because of the foliage of the bushes.

For subdivisions of the officinal valerian group, see the Appendix.

Valerian often becomes visible again above the leaf canopy of the under-growth along with the flowers of wild privet (*Ligustrum vulgare*), elder (*Sambucus nigra*) and others. Although such sites are sometimes damp tending to boggy, and valerian grows in communities with meadowsweet (*Filipendula ulmaria*), stinging nettle, herb bennet, and common horsetail, there are also dry to rocky sites where it thrives. An extreme example is provided by a Swedish island called 'Scent Island', whose craggy rocks are completely overgrown by valerian.

In Scandinavia, valerian is also found on decomposing material deposited at sea margins. The predominant climate there is cool and humid with short days at the beginning of the year and long summer days with a brilliance that is enhanced by reflection from the surface of the water. It gives rise to sturdy plants doused in red with a superabundance of flowers, but in the tropics it is difficult to get true valerian to flower.

Plate 7

Yarrow, midsummer, patchy pasture

On this unfertilized pasture in Sweden, groups of yarrow in many shapes and sizes are in harmony with their background. The green of the plant's fine structures appears much darker than that of chamomile.

At the time of yarrow's flowering, from June to autumn and often even in winter, the vegetation is much more at rest, unlike that in the picture of the dandelion surroundings. It has already begun to ripen and to become more rigid. Most of the meadow flowers and grasses have already faded.

A yarrow mood also arises at very barren sites, such as unfertilized pastures where everything is grazed short, and sites that are very dry. There it forms a short carpet over the ground, with flower shoots only a few centimetres tall. At such a site the dark green leaves are narrow and pointed like an awl, and so dense that the leaflets of their finely pinnate leaves can be distinguished only at close range. Thus, in relation to the greenness of the plants, the flowers are much more striking. By contrast, at a somewhat shadier site, where fertile arable land borders a ditch, yarrow reaches well above the knees, the leaves broaden and here the leaflets appear like a fine, open network. Even so, in situations of enhanced growth, the outline of the overall leaf shape remains clearly discernible.

Yarrow usually does not grow in the dense vegetation of luxuriant hay meadows, but it can spread and sustain itself where fallow arable ground gradually merges with them. There it contracts more towards the ground while often its flowers only appear with the second season's growth. Thus, yarrow presents itself in an atmosphere of patchy fertile meadows and pastures, arable fields and roadsides. Its whitish parasol-shaped flowers are also found at sites where red, ripe, sweet wild strawberries (*Fragaria vesca*) are perhaps thriving.

Plate 8

Oak, in summer, open oak woods

Although oak flowers in May before peak leaf growth, it is impressive not for its flowers but for its massive, gnarled presence. The open, green space it creates comes into leaf late in the year, and when in full leaf it also lets light through so that other trees like hornbeam (*Carpinus betulus*) and many shrubs can develop underneath, as shown in the picture. Oak woods or clumps of oak trees always attract a wealth of plant and animal life. Buds and leaves attract many species of gall-forming insects; rotting wood attracts stag-beetles (*Lucanus cervus*); the resinous exudate of young bark attracts butterflies and other insects; acorns attract jays (*Garrulus glandarius*), wild boar and deer.

Plate 9

Common horsetail, in summer, open land with moist soil

In the cool atmosphere of an early summer morning we come upon a clump of horsetail growing by itself in a moist open location. A wonderfully rhythmic pattern of countless, delicately glittering water droplets hang like rows of pearls on the tips of its slender shoots, and shine in the sun over the whole area. Later in the day, the site appears finely structured, as in Plate 9, and now reveals the atmosphere that enables horsetail to form such a crystalline structure out of the moist subsoil.

Plate 10

Summary

The illustrations of the preparation plants' habitats at the peak of their development are intended to convey an initial and pictorial (imaginative) impression of the atmospheres that enable the various species to develop. In practice the places where they grow can appear different from the examples shown. As a result, it is easier to see from the variations in the plants' shape, what qualities in the surroundings influence them in particular. The impressions of the habitats are initial experiences of the overall conditions for the preparation plants to flourish, and are thus experiences of invisible organs in the landscape that comprise the particular environmental effects. These illustrations are all the more revealing if, in contemplating them, we consciously resonate what is of course not at present visible in them, but, like the course of the day and year, is an essential complement to them. Just as we need our knowledge of the zodiac and of the constellations at a particular moment in time to understand the position of the sun on a given day, or just as we need consciously to associate the appearance of a plant with its annual development and that of the place it grows if we want to get nearer its essential character, so too is it necessary with atmosphere to visualize what takes place in time but is experienced as spatial. Immersing ourselves in these processes means at the same time recognizing our own relationship with them.

For stinging nettle in light and shade see, page 22f.

4. The Year as an Archetype of Human Soul-activity

Our thinking is a light process. One of the most important exercises for immersing ourselves in the contexts of the earth's life comprises observing the thinking process and its alternations between the mechanical and holistic faculties of perception in harmony with the seasons. We learn to recognize the pendulum movements of our own soul so that with them we can turn our contemplation to the sense world and out of it work with the forces that manifest their effects in the cosmos. This plays an important part in understanding the preparation plants, as well as in producing the biodynamic preparations when — as we will see later — we expose them to the influences of the seasons by hanging them out in the air or by burying them in the ground.

Rudolf Steiner published a soul calendar for people wanting to be wide awake to experiences and changes over the course of the year. It contains a verse for each week of the year that expresses the relationship of our own soul to the cosmos at each particular time. Rudolf Steiner wrote the following in his foreword to the first edition of the calendar in 1912/13 :

> As human beings we feel united with the world and its temporal
> changes. We find the likeness of the world's archetypal image in our
> own being. This likeness is no sensory or pedantic imitation of the
> archetype. What the great world reveals in its temporal flow
> corresponds to a pendulum swing in our being that does not move in
> the element of time. Our sensory and perceptual being, we feel,
> corresponds much more to the nature of summer, woven through with
> light and warmth. During winter's existence, we sense ourselves much
> more grounded in ourselves and living in our own thought and will
> worlds. Thus, the rhythm of inner and outer becomes for us what
> nature in its temporal alternation presents as summer and winter. A
> great mystery of existence can rise if we bring our timeless rhythms of

Rudolf Steiner, preface to the first edition of his 'Calendar of the Soul.' 'Anthroposophischer Seelenkalender' (1912/13) in Wahrspruchworte, *Dornach, 1991 (GA 40); trans:* Calendar 1912/13 *(Great Barrington: Steiner Books, 2003).*

perception and thought into correspondence with nature's temporal rhythms. If we do so, the year becomes the archetype of human soul-activity and thus a fruitful source of true self-knowledge.

With this in mind Rudolf Steiner allowed the Soul Calendar verses to be included on cigarette packets as a support for soldiers at the front line.

A few examples of the Soul Calendar verses illustrate what it is about. The contrast between summer and winter is expressed in the following verses:

> *Summer*
> Fourteenth week (7–13 July)
>
> Surrendered to the revelation of the senses
> I lost the drive of my own being,
> And dream-like thinking seemed
> To numb and rob me of my self.
> But cosmic thinking in sense appearances
> Is approaching, awakening me.
>
> *Winter*
> Thirty-seventh week (15–21 December)
>
> With joy my heart's desire strives
> To carry spirit light into the winter night of the world,
> So that shining seeds of soul
> May root in world foundations
> And the word of God resound in senses' darkness,
> Illuminating all existence.

The autumn verse draws our attention to how summer impressions are taken up into the soul, transform themselves and stimulate our self-reflection.

Autumn
Twenty-seventh week (6–12 October)

When penetrating to the depths of my being,
Expectant longing urges me
To find myself in self-reflection
As a gift of the summer sun, as a seed
That warming lives in autumnal mood
As driving force of soul.

In contrast, the spring verse draws our attention to the way in which, in the early part of the year, our soul once again turns outwards to what comes to it from the periphery through the world of appearances.

Spring
Easter mood (7-13 April)

When from cosmic expanses
The sun speaks to human minds
And joy from the depths of soul
Unites with light in seeing,
Then, from the sheath of self-hood,
Thoughts ascend to distances of space
And weakly bind
The being of man to the spirit existence.

By working regularly with the verses we gradually develop a sense for the subtler transitions from one season to another. Just two examples will serve here to illustrate these transitions.

The mood characterized by the spring verse is prepared step by step out of winter, for example:

Forty-eighth week (2–8 March)

Let certainty of cosmic thinking
Appear in the light
That would stream from cosmic heights
With power toward the soul,
Solving the soul's enigma,
And, focusing its powerful rays,
Awaken love in human hearts.

After the spring equinox our experience is guided further towards summer:

Sixth week (12–18 May)

My self has arisen
From the bounds of my individuality
And finds itself as cosmic revelation
In the forces of time and space;
The world shows to me everywhere
The truth of my own likeness
As the divine archetype.

Then the verses make the transition to autumn:

Nineteenth week (11–17 August)

Mysteriously to enclose with memory
What I have newly received
Will be my striving's deeper goal:
Strengthening, it should awaken
The inner forces of my self
And, in becoming, give me to myself.

And after the autumn equinox the mood turns towards preparation for winter:

Thirty-fourth week (24–30 November)

Mysteriously to feel within
What I have long retained
Reviving with new found self-hood:
Awakening, it should pour
The forces of the cosmos into my outer deeds in life
And, in becoming, shape me into existence.

Many other relationships, including those of structural composition, can be discovered amongst the verses and used as objects of contemplation. This can enhance the richness of our soul-experience over the course of the year, and further awaken a sense for the connections between the concrete phenomena around us. Each person will find their individual approach, which will of course already be modified by the circumstances in the part of the world where they live. It is precisely the individual approach that forms the basis of fruitful participation in the healthy development of the earth. There is nothing compulsory in these exercises. On the contrary, each person must make and trace out their own inner connection to the phenomena of the senses. And this involves giving our attention to real events.

Aspects of the cosmos. Four orientations that complement one another

By 'cosmos' we have in mind its original meaning as the spirit spread throughout the world, in the sense of universe, beauty, order and radiance or adornment. These are various aspects that we view here as orientations. They can prepare the way for encountering the spiritual inherent in the phenomenal.

Universe or *macrocosm* is usually understood as all-encompassing, boundless. If we seek this totality in the manifold phenomena, it is already at work in our willingness to turn to it. We have already grasped it intuitively as our line of enquiry, but at the same time as an open question. Universe also lives in our intention to turn to a particular being. In this process we grasp the significance of the microcosm in relation to the macrocosm as a relationship of complementarity between the world and our own inner being. In this view, every being appears as an infinitely extendable world, as the whole in the part. Thus the universe is embodied in every question about the essence of a phenomenon that has our attention.

Beauty

We apply our aesthetic sense to experiencing beauty. We look at a crystal, plant or landscape in such a way that we try to find in them the completeness that manifests itself in the appearance but, in addition, points to the spirit.

That means immersing ourselves in the experience of the phenomenon, becoming one with it. Percept becomes picture; becomes the inner experience of a being that speaks to me just as long as I can remain in it, alert and open.

In this experience we are exposed to two temptations. One diverts our attention from experience of the other being. Our soul wants to merge with the other or 'wallow' in what is sympathetic in it, at least to the extent that we are no longer fully present. The other attempts to make us inclined to latch onto something 'tangible' and put this in place of observing. This temptation corresponds to a longing to be able to describe a particular outcome of immersion in our feelings. however, the real outcome of this engagement with the other comprises the developing *faculty* of lovingly connecting the situation consciously with certain beings again and again in life, and gradually learning to look at things from their point of view.

Order

We can easily lose ourselves in pure perception. However, the search for principles of order can be assisted by the experience of beauty as it is this that organizes perception. We then learn to proceed with our inner activity in a particular way from one part to the next, while conscious of the whole, and to trace or track our movements on this journey in thinking. This leads to experiencing in *pure thinking* the cosmic that is connected to the *way the being manifests itself* in earthly space, for example in the kingdoms of nature. These tracks in thinking, the ordering relationships, are particularly suited to pure mathematical-geometric contemplation. Science generally uses it in an elementary way, but through contemplating our own thinking activity this faculty can consciously become an organ for perceiving the essential nature of the ordering relationships between the qualities perceived.

We can also see an element of beauty in order, but then we are moving in the generality of principles of relationships. *What* is related to something else is, however, immediately accessible to experience through pictorial perception via the sense of beauty.

Radiance or adornment

The three previously characterized outlooks on the cosmic coincide in the concrete reality of the particular here and now. Radiance or adornment, considered as an outlook in the sense meant here, means on one hand, a first encounter that awakens our interest and on the other, the *quality* of what we have before us that we have grasped after a longer journey of contemplation, and also from the essence of the thing.

5. The Connection Between Plant Life and our Own Experience

If we have ever been deceived by an artificial flower, the question might arise as to how it comes about that we can immediately recognize a plant as such. What expectation do we unconsciously add to the percept?

From germination, growth, flowering, fruiting and fading, etc, everyone probably has ideas, albeit imprecise, about the life of a plant. These they have developed in the course of their life into a fundamental outlook. We immediately perceive plants with this outlook without being aware of doing so. Without such a specific capacity to observe we would see them only as some unspecified thing. This outlook is the germ of an organ of perception for the essential nature of plant life and can be developed further through regular observation.

At each developmental stage of a plant that we are studying we begin to develop a different relationship. Inner complementation of every individual plant phenomenon resonates with what we see, in a manner that is the same as the delicately yellow shining after-image that forms in us after staring at a blue colour. With a change of phenomenon in form, colour, smell, etc., this inner complementation also changes according to the given manifestation of the plant.

For after-images, see page 16.

For example, when we sow some inconspicuous seeds, our expectations are associated with a picture of something gradually appearing and filling itself out materially. If we do not know the identity of the seeds, our initial expectations are of only a very general nature, but things become clearer as they grow. We see details, characteristics of the particular plant, but we also become aware of inadequacies in the environment that we provided at sowing, such as the quality of the soil and lighting, and of our subsequent care. In extreme cases, we do not discover which plant species we are dealing with until it flowers.

However, our relationship to a wilting plant is different. It is often difficult to reconstruct a memory of the appearance of the flowering, mature plant from the deformed and discoloured shape that remains. But greater effort of memory allows it to live on in us. If fruit and seeds have been produced, then our relationship to the plant changes again.

Our thinking and perceiving begin to move if we proceed from the picture of a phenomenon that we have perceived and retraced in our mind's eye, to actively continuing in the direction from which this picture emerged and in which it leads further. Individual thoughts disappear. We begin to experience the effects of forces in our soul and grasp them with perceptive thinking in the changes that give rise to the forms of the living world. These forces should not be confused with mechanical forces. In passing from one perceptible stage (that to our senses appears as something that has come into existence) to the next, and to retracing and experiencing a formative process, we also notice changes of a different kind in which specific rhythms are revealed.

What comes to life in us through following the plant world in this way is best pursued in the course of the central European seasons from winter to spring, summer, autumn and back to winter. A relationship between our soul-life and plant life with earth and cosmos is discernible in what lives in us from the seasons as a more general rhythm of the earth's life. The exercises with the soul calendar are good for attuning ourselves to this. Every plant species lives a one-sided reflection of what, in the general course of the year, stands in a particular relationship at any one time to the organism of the earth. This can strike a chord in us and enable us to grasp from inside what is special about the life of a plant species. Awareness of this inner vision can be awakened by connecting what we have observed step by step in a plant with following the seasons in conscious experience. This brings about an elementary foundation for understanding the special character of individual plants and their thera-peutic effects.

A plant that can complete its development several times in the course of a single year — groundsel (Senecio vulgaris)

In the pictures of each individual plant we can see the result of development over eight weeks at a particular time of year. The sequence yields a formative gesture that is typical of years in Central Europe. Spring and autumn clearly show the contrast between the parts of the year towards and away from the summer solstice:

February–March: The shooting upwards of the greenery, away from the ground, predominates. The root connects with the soil in fine branching.

October–November: The plant above ground stays close to the surface of the soil, whereas the root penetrates deeply into the ground.

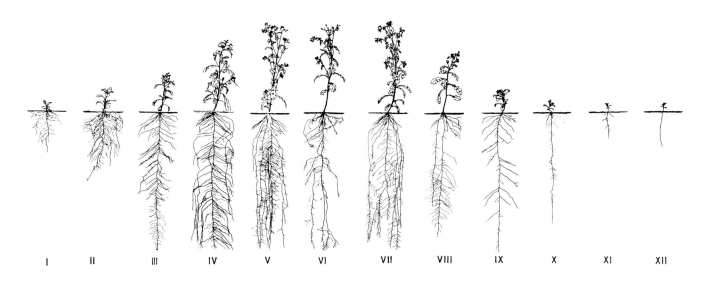

I   II   III   IV   V   VI   VII   VIII   IX   X   XI   XII

From: Jochen Bockemühl, Lebenszusammenhänge erkennen, erleben, gestalten, *Dornach, 1980; trans.* In Partnership With Nature, *Jochen Bockemühl (ed.), Biodynamic Literature, Wyoming, Rhode Island, 1981*

Contrasting ways of thinking for understanding the phenomena of the kingdoms of nature

Mineral form and function should be understood in a different way from that of plants, animals and human beings. We recognize such differences in that the particular forms and functions prompt us to pursue specific ways of thinking. For example, when we look at a *crystal*, elementary geometric thinking is adequate for understanding its ideal form as a whole in space, against the background of all the possibilities of spatial manifestation. We do justice to phenomena connected with particular substances if we follow a seed's disappearance without trace as it dissolves, and its re-emergence in a new, concrete crystalline form.

In contrast to crystals, plant forms manifest themselves only as intermediates in a process that we must follow in order to experience the whole. Plants are always open in relation to the environment. Over time, our pictorial, imaginative perception allows the whole that is active in the appearance of a plant to arise, guided by our increasingly clear way of thinking about it.

Applying geometry to *plant forms* after the fact (searching for numerical relationships, leaf shapes, spirals, etc.) can no doubt enthuse us, but in the end it remains unsatisfying because it cannot do justice to the process in change, for example between the pictorial phenomenon of a plant that has culminated in a flower and a seed that, as a tiny grain, is connected with the entire compass of possible phenomenal forms. Both stages show something that is roughly understandable geometrically, but the transitions remain incomprehensible because in this case we would have to imagine an inversion in the formative process.

Projective geometry can help us trace the movements in thinking, that enable us to understand the different intermediate manifestations of the phenomenon, if we follow the necessary movements with complete openness. A precondition is that, through perceiving with our sense of beauty, we already have a first inkling of the being we are approaching. Thus understanding the transition from seed to flowering plant and back into seed requires the control of formative movements in thinking that trace out a guided trail of transitions from space into counterspace and back into space, in order to deal properly with the transition of manifestations of sequential forms (metamorphosis). This movement in thinking should not just be worked out, but rather permeated with the experience of the whole human being. Flowering plants and seeds are in a *relationship of inversion* to one another in that the plant expresses its form outwardly and the seed, by contrast, inwardly. The two polar opposite pictures of a plant can be seen as two sheets of paper one showing a credit and the other a debit. Both of them appear to the senses but our relationship to each is totally different. Understanding the transition between one and the other requires that we trace the appropriate route with our thinking. If, having traced the particular route, we return to aesthetic perception of the phenomenon it becomes much more consciously differentiated and richer.

By contrast, *animal forms* always appear as a complex whole that is differentiated during development. We perceive animals by immersing ourselves in their shapes and movements with our sense for beauty, and experiencing them as specific ensouled beings. Guided by the souls that are active in them, they unfold their lives in space. There they are attracted or repelled by forms relevant to them or they create such forms themselves.

This invites the discovery of other mathematical principles, that is, ways of thinking.

Here these will only be indicated. The relationships of perceptual experience to mathematical constructs are so far little researched in relation to the nature of animals and man. The approach described here provides a valuable tool for accessing the principles of the plant world. It helps to bring clarity into the training of higher cognitive faculties that accompany perception, in the sense of perception of beauty, and to make it understandable. For further material on this see Rudolf Steiner, *Das Verhältnis der verschiedenen naturwissenschaftlichen Gebiete zur Astronomie* (GA 323), Dornach: Rudolf Steiner Verlag, 2nd edition 1983. (Translated by George Adams, typescript available only).

6. Experiencing the Course of the Year as a Whole

People who, like farmers or horticulturists, adjust their work according to changes in the natural world, experience the seasons as changes in their own work. Townspeople often lack such a connection. Their relationship to nature can easily become superficial and sentimental.

Unfortunately, horticultural work and mechanized farming have largely become routine. Thus the most essential features of life are easily overlooked in them. People increasingly think in terms of technical concepts and leave their feelings to themselves, because they are no longer related to the world of directly perceived phenomena.

We do not know what relating to the course of the seasons meant in earlier times. Even so, we can imagine it. But it is pointless longing to be back in a time when people felt determined by the starry heavens. Consciousness and casts of mind change. But what can now replace the direct contact we have lost with our natural surroundings? How can we create it in a new way? We notice in the following extract how Rudolf Steiner described the reversed relationship of the modern human being to the stars in that he took his starting point as the actual experience of it at a particular moment.

Explanation

An extract from the foreword of Rudolf Steiner's *Kalendar 1912/13* from *Beiträge zur Rudolf Steiner Gesamtausgabe* 37/38, Dornach: Rudolf Steiner Verlag, Spring/Summer 1972.

We experience time through changes in worldly phenomena. These changes link new with old as the world goes on its course. Day follows night. Night follows day. The new day does not allow what has passed to arise again from the womb of existence, but it repeats the previous day in its own essential form.

The light of the moon shines into the darkness of night. It waxes over fourteen days and nights then wanes again for the same length of time. This too constantly repeats itself, the new preserving the old.

Through the power of the sun, plant life is enticed out of the ground. It unfolds, wilts, retreats back into the concealment of the ground, like day into night, or the brightness of full moon into the darkness of new moon, and arises again; the essence of the old always revealed in the new.

Human beings are confronted with this world-becoming which changes and in changing preserves itself. They have to harmonize their own experiencing with this cosmic activity ...

The repetition of the old in the new expresses itself most characteristically in the relationship of the stars with one another. These positions constantly recur, to the extent that the new ones are like the old. People can express the relationship of their experience to a particular point in time by referring to the position of the stars at that time ...

We take a look at one of the pages in the following calendar. We choose a particular day, for example in May, and then another in August. All the experiences that we have on these two days in our involvement in world-becoming are totally different. We can express these differences by, for example, relating to our experience the position of the sun in a constellation of the zodiac, as with a letter of the alphabet to the sound it represents.

Experiences in the course of the day

The course of a day reflects in miniature what happens in the course of a year. Through perception of our own self, in thinking, feeling and willing, we experience our existence as continuous. But in fact it follows a rhythmic course, like breathing, changing with our orientation outwards or inwards, only we do not usually pay attention to the inward direction. However, this inward direction is extremely important for understanding our relationship to the essential nature of the world and thus for our own conscious involvement in earthly processes.

In day consciousness we are mainly self-consciously purposeful, awake, clear, discriminating. In sleep we are not conscious of our self. We sink into sleep with what we have absorbed through ordinary thinking during the day. With the right attentiveness, on the following day we might notice that we have altered a bit. We have a different approach to something we see again, perceive it differently from the way we did the day before. It is worthwhile giving special attention to this process.

Living in the rhythm between the sense world and the cosmic world as an inversion process

There are several different observations which make us aware of the way that the aforementioned alteration in us happens as a kind of inversion process in the change between waking and sleeping.

For the principle of inversion, see page 65.

In the day we live in consciousness in the sense world that surrounds us. We form ideas about individual appearances and objects, of how close they are to each other and the way in which they follow one another, and we adjust our outwardly directed actions accordingly.

These ideas often continue to be reflected dreams, but they gradually sever their direct connection with what was perceived by the senses, grow somewhat, acquire different circumstances whose contexts seem puzzling. These grow in incomprehensibility and with this we sink into deep sleep as we lose consciousness in the infinitude of scenes. Then in our sleeping life, we manage in an elementary way to extend our thought-experience into the self-contained cosmos of the spirit world. We participate in the incomprehensibility of this spirit world with our perceptive thinking, which is normally dormant. We could also describe it as thinking from the periphery, in contrast to mechanical, objective thinking. What continues to resound from it in the soul during the day, for instance at the sight of a plant, enters our daytime, conceptual thinking and doing unconsciously.

Frequently, we view differently things that we were involved in during the previous day. Often we conceal this opportunity with our emerging day consciousness and carry on in the usual routine of received ideas. At opportune moments we notice how, for example, this at first lives dreamily in our holistic-aesthetic perception of a landscape when we awaken an insight into the connections that we could not make on the previous day when looking at individual objects. A question goes with us into sleep and inverts itself during the night into an organ of perception for the particular connections that escaped us the day before.

However, this insight can slip away from us again in the course of the day. Something about it perhaps appears trivial to us, and we feel deceived. But it often proves worthwhile too by invigorating our activities and giving them a new direction. We gain new impulses to do something that fits better with the context of life.

Over time, in our daily life, this leads to the formation of yet unconscious germs of organs of perception for organic life that appears different in each situation. This faculty of perception can be developed with exercises.

Observed more closely, our consciousness changes in rapid alternation, even in the daytime. We involve ourselves with a percept, stand back from it again in thinking, and unconsciously connect ourselves with the complementation that has taken place and lives on in us. But we are usually unaware of this. We speak glibly of experiences without heeding the most important aspect, namely what kind of experiences they were.

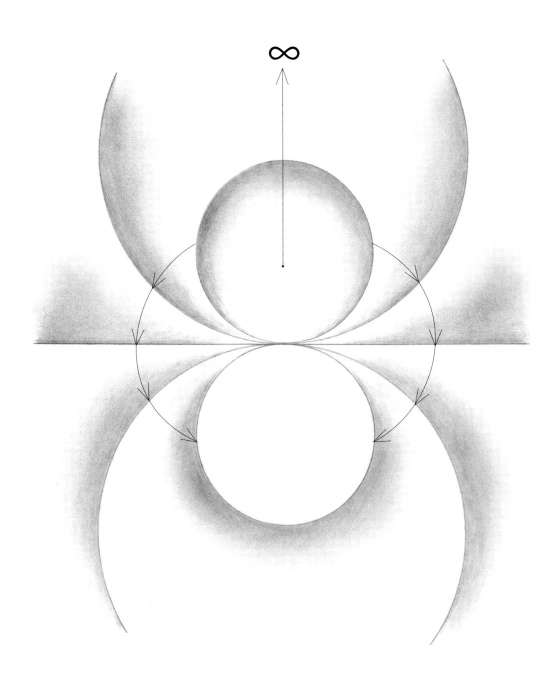

The principle of inversion in the sense meant here

Here is an exercise in separating thinking and experiencing from a spatial context in such a way that they begin to move and open up to spiritual perception. We picture a circle that is growing larger and larger and we keep its point of contact with a fixed horizontal tangent. The curvature of the circle thus decreases. Soon it becomes a vast ever-broadening circle that is no longer conceivable to the extent that we could follow it in the ordinary way with our mind's eye. It finally becomes an infinitely large circle and the part of it that we can illustrate on the page becomes a straight line that disappears to infinity in both directions and coincides with the horizontal tangent. Our ordinary thinking based on mental pictures increasingly comes up against contradictions. Can a circle be a straight line?! But if we do not stop thinking, which of course continues on in the purely spiritual element and is not bound to mental imaging, then as we pursue the process we awaken through our thought activity to totally new percepts, for example, that the circle that became a straight line is closed through infinity. Infinity is present everywhere; it is not spatial.

Something similar happens to the central point of the transforming circle. It is a point that is equidistant from every point on the circumference of the circle. If we follow it as the circle enlarges further and further away along a line vertical to the straight line, our spatial thinking must once again change into a movement in thinking directed towards infinity.

There the centre point takes on completely new characteristics, which are of course unequivocal, but can only be taken into ordinary thinking through pictorial clues. It has now acquired the features of a 'spaceless straight line'.

If we continue the movement through infinity, the circumference of our circle that became a straight line begins to curve in the opposite direction until finally a circle reappears which is outwardly identical with the first only 'inverted' with respect to it. We can appreciate this through the way we brought it into existence. The centre of this inverted circle *is new*. But what became of the original centre? Putting it in pictures we could say that on its further journey it has 'bent' so much that it has become the *infinitely distant periphery*, that constitutes from outside, out of the spacelessness of the infinite, the new visible circle reappearing in between.

But that can be understood only with purely spiritual perception. It is a reality that is presupposed unconsciously when anyone refers to holistic etheric effects. What thus results from contemplation of thinking can be saturated meditatively with experience. At first it is a formal picture. This can acquire content when we recognize in it traces of the process that we go through with our consciousness as we enter sleep — that is, in the transition from day to night consciousness. If we go out of space in this way we at first lose waking consciousness based on mental picturing. And when we wake up again, our consciousness has taken on a reversed relationship to things in space. In this inversion process, what is apprehended mechanically via perception 'without context,' changes in our soul into a kind of spiritual organ of perception. On the following day, we use this organ to view anew, as if from the infinitely distant periphery, the phenomena of the previous day in their overall relationship with other phenomena living in our consciousness. The process of geometric thinking can help us experience the activity of feeling our way to spiritual reality without losing the ground under our feet by doing so, in other words in a rudimentary way while remaining present with self-perception.

The course of the year as a life process

For approaches in thinking to understanding the phenomena of the natural world, see page 58.

For fundamental experiences of a comprehensive pictorial-imaginative understanding, see page 115.

In the broader rhythm of the course of a year, something happens that is similar to what we have described for the course of a day.

The examples of landscapes, shown in Chapter 3 (pages 20 to 48), with the help of specific situations, reveal the way in which we relate differently to landscape according to the change in phenomena at various seasons (for example with dandelion and yarrow).

The changes in vegetation make visible the life processes through which we are involved in the life of the earth. Our senses are presented only with a particular state of things each time. But if we perform the aforementioned transformation of one thing into another with attentive observation attuned by the Soul Calendar, we will learn gradually to understand not only the forces that produce vegetation, but also those that make it disintegrate outwardly while new formative potential concentrates within. We experience these forces as real and active in the living world, but differently from the way we experience physical-mechanical forces. By following the change in vegetation in this way, the process becomes one in the midst of which we ourselves are placed.

We are always in the present with our perception through all our senses, but, with the aforementioned attentiveness, we perceive in the present picture of the phenomenon of a plant its earlier developmental states; we look at it and experience it as a living whole.

At the level of consciousness, we have made the spiritual aspect of a plant so much our own that it is like a seed which, in the process of nature, acquires the potential to form new plants. In the case of the plant a new receptivity to environmental influences arises in the present. In our own case a deepening of receptivity to new percepts arises.

Just as light and soil conditions affect a plant form in different ways according to whether it germinates or has a tendency to flower, so too our receptivity changes for new percepts in accordance with what has reorganized itself into perceptive capacity in us from previous sensorial encounters — that is, the preparedness that we bring with us when we have a particular intention in mind. At the same time it is, of course, important to develop with our imagination many mental pictures of possible realizations of form, so-called models. But these will only serve to invigorate the germinal force of the idea in consciousness if such pictures, which are prone to hardening, are dissolved in the process of becoming. This is the situation with the germinal shoot in seed and bud or in the earth in winter.

If the model is solely like an inner indicator — that is, taken spiritually — then it facilitates action based on observation with presence of mind.

In winter, plant life is least conspicuous and yet at that time it is at its most vital. Subject to the limitations of the species, all possibilities for reappearance remain open in seeds and buds, in contrast to which any subsequent emergent growth already signifies becoming one-sided, that is, limited. While outdoors we see living nature asleep; seeds in the earth and buds are of course going through a period of dormancy. Thus we should not take cuttings for rooting before St. Barbara's Day (December 4) because the vegetation must first reach a certain point of development inside. But after that point, in the depths of winter, we notice minute changes and sense that something is stirring within, pressing to outer activity. As the days get gradually warmer, snowdrops (*Galanthus nivalis*) and goldilocks buttercups (*Ranunculus auricomus*) begin to emerge from the ground and hellebores come into flower.

The transition to spring, summer, autumn and to winter awaken in us specific sensations. We can connect this consciously with what emerges from invisibility into visibility in the plant world, then loses mobility and withdraws again from the perceptible world through fading and wilting. This activity can create a vivid picture if we repeatedly appreciate the course of the year as a breathing process. In Central Europe the balance in the alternation between the outward trend and the inward is at its most pronounced and most harmonious.

This reality can yield new content for the annual festivals. They will then no longer be celebrated purely out of a recognition of tradition, but people themselves will give them a new meaning that makes them more conscious of these processes.

Order among the stars and beauty on earth

In the cosmic element of the universe, order among the stars and beauty on earth interpenetrate one another.

Differences can only be found in them by changing our point of view. At the same time we need to be very conscious of it if we want to penetrate the supersensible spiritual realm, because in that realm things are not positioned separately, one beside another, but are connected and interpenetrate one another in the way they do in our world of thinking and experience. Where we choose to turn our gaze is always an essential part of contemplation.

We encounter the cosmic in two ways when we look at the sky and contemplate the relationship of what we experience there to the active forces. On a sunny and cloudless day, in the sensation of the blue of the sky, we are looking at the cosmic-etheric element. In the night, the stars appear as points of light that we normally regard as external objects, or we associate them with unspecific experiences. Or again we notice constellations that we can distinguish from the arrangements of the stars, name them and later recognize them. Pictures arise *in us* if we are really thoughtful and observant when we study the stars and their ordering principles. These inner pictures are, of course, not the mental images that, for example, we reproduce symbolically on star maps. They are soul-spiritual percepts, *ideas*, that certainly cannot be pictured in the mind's eye but are ways of viewing the world that, at first, point our spirit gaze in various directions. Our contemplation is referred by these soul-spiritual directions to the cosmic-astral influences.

Principles in the constellations and the movements of the stars that call for mathematical-geometric thinking, form above all the aspect of cosmic *order*. However, during the day when a variety of phenomena in the sense world around are impressing themselves on us, we encounter *beauty* through rich pictures in colours, scents and structures. These pictures call for our aesthetic sense and enable us to perceive the archetypal form that we are immersed in with our soul. Order and beauty complement one another. One cannot be reduced to the other.

Conventional science bases its power of conviction on the ordering relationships between qualities by aiming to convert observations into numerical and measurable form. This overlooks the fact that, for example, qualitative investigations presuppose the quality as a given. But such a presupposition requires at the same time that we think about what we actually experience in the broadest sense as being connected with beauty, harmony and image of a being. However, this self-reflection is all too often replaced by opinions, received ideas, etc., about, for example, what 'beautiful', 'healthy' and 'vigorous' should mean.

The sun by day illuminates the phenomena of our natural surroundings, and, complementing this, night enables us to see the world of the stars where, against a background of the fixed stars, sunlight is reflected and modified by moon and planets. So too, summer, with its fullness of plants and insects, opens the richest view of the earth to our sense of beauty, whereas winter, by contrast, stimulates us to engage spiritually with the cosmic element connected with the earth, as with a seed. In tuning in to such experiences, we participate more deeply in the cosmic process.

If we then go into the way in which order and beauty permeate each phenomenon that we want to form a judgement about, we are led to insights into the reality of the quality of the hear and now.

No season is experienced in isolation from the others. They are all spiritually active in the background. It is nonsense to speak of winter without presupposing the other seasons. In experiencing the current situation (for example flowering dandelion in spring) we bear in us the image of the snow-covered meadow and develop certain expectations for its further development in the summer. Experiencing the course of the year occurs through supplementing, in inner perception, the present with the past and with a preview of what is to come.

The more deeply we penetrate the experience of the course of the year, the clearer becomes the basis on which all natural phenomena speak. Theirs is the language that can guide us in all our endeavours, not only in those of agriculture.

Thus, our soul expands in summer when the sun has risen to shining brilliance in the divine process of exhalation of light, united with the widths of the world. The light speaks and human beings can discover their spirit origin in the phenomena. In the mystery of summer, the natural world is the image of the night, as illustrated in different ways by Novalis in *Hymns to the Night* and Shakespeare in *A Midsummer Night's Dream.* What appears to the senses is a picture of processes in our own experience, and vice versa.

In the breathing process of light during the year, the earth sleeps soul-spiritually into the summer. The light of summer creates sense perceptible pictures, but our experience of winter leads us into relative picturelessness. When the living world appears to be asleep, the earth is spiritually awake with its preparation processes for new growth. The sun now works spiritually.

People take their summer experience into winter and there wait for something to be born from it. At this time the Virgin bearing the Child is queen. With this image, our soul questions its own moral nature. Will it be fit enough to bear this fruit? It is reminded to become worthy of its spirit garments. This is a matter of will.

The significance of pure sense experience and pure thinking

It was not until the age of thirty-five that Rudolf Steiner, after writing at Weimar about Goethe's approach to cognition, realized the significance of *our own* sense perception for knowledge of the spirit. He wrote about it in his autobiography:

> It was as if I had not been able to pour the soul's experience deeply enough into the sense-organs to bring the soul into union with the full content of what was experienced *by the senses.*

This new discovery encouraged him to devote himself more intensely to sense perception than hitherto:

> For me the enhancement and deepening of the powers of sense-observation meant that I was given access to an entirely new world. The placing of oneself objectively over against the sense-world, quite free from everything subjective in the soul, revealed something concerning which a spiritual perception had nothing to say. But this also cast its light back upon the world of spirit. For, while the sense-world revealed its being through the very act of sense-perception, there was thus present to knowledge the opposite pole also, to enable one to appreciate the spiritual in the fullness of its own character unmingled with the physical. ... I soon learned that such an observation of the world leads truly into the world of spirit. In observing the physical world one goes quite outside oneself, and just by reason of this one comes again, with an intensified capacity for spiritual observation, into the spiritual world.

Rudolf Steiner, *Mein Lebensgang* (GA 28), Dornach: Rudolf Steiner Verlag, 1999; trans. *Autobiography*, Part XXII, (Anthroposophic Press: New York, 1999.)

7. Formative Processes of Individual Plants

Each plant takes shape not only as an image of environmental conditions that inform its developmental potential (site, seasons) but it also reflects the course of the year in germinating, leafing (greening), flowering, fruiting and wilting. After germinating it gradually produces leaves. If we arrange the finished leaves in a sequence side by side according to their order of appearance, the resulting series of leaf shapes enables us to see a change that is not *physically* perceptible. It shows us the leaf at first separating from the shoot by lengthening its petiole and its blade gradually spreading. Then, more often than not, it develops an increasingly complex and deeper division through indentations that frequently culminate after the greatest extension of the leaf on the way to the flower. The leaf tips protrude more visibly at the periphery as what we may call the centre of gravity of the leaf moves closer to the plant stem, its petioles shorten, become more integrated into the leaf area, often with the leaf base or secondary leaves extending significantly. This process, one that does not take its course physically, is perceived so concretely in inner contemplation that we are often led to think that we can see it with our physical eyes.

Seen this way, the progression of forms is already relatively complete that

Leaf sequence of coriander (selection)

embraces flowering, fruiting and fading. It is organized in many different ways in the various plant families. Together with the way the flowers and fruit are formed and how the plant fades, it reveals specific gestures for each plant species to inner perception. This development of form has a particular relationship with the general course of the sun throughout the year. It can have a different relationship with the seasons, comparable with the movements of the planets. Quick growing plants can complete their cycle of growth to the seed stage several times in a single year. This is analogous to the moon's passage through the night sky. Trees, by contrast, span several years with their growth cycle. In the zodiac, this corresponds to the movement of the planets above the sun.

The cosmic influences conveyed by the sun over the course of the year and the predisposition to *receptivity* for these solar effects interact with each other in the growing and fading of plants. These effects are modified by the conditions on earth through the seasons, while moon and planets take in and radiate back the light of the sun in manifold ways, modifying the solar effects. Furthermore, during the general course of the year, development reaches its peak of maximum unfolding in each case in a different relationship to the summer solstice.

If we take it for granted that in our latitudes — Central Europe, 46–57°N — the flowering of the vegetation increases mainly towards summer, germination occurs in spring and fading and fruiting in autumn, then we should ask why a plant such as spring crocus flowers in spring, that is, before the greatest growth, or autumn crocus (*Crocus nudiflorus*) in the autumn, when vegetation has largely faded and ripening has begun.

The *atmospheric aspect* of the environmental effects is expressed in the special individual form to the extent that it points to shady — light, infertile — fertile, seasonal rhythm etc.

The specific gesture of each species shows their particular *cosmic alignment* that distinguishes them from the general background of the language of gesture of the vegetation, as do constellations *vis-à-vis* the fixed stars. Through combining all qualities as a whole in the phenomenon in the aforementioned sense, the cosmic alignment of a given plant is expressed through its particular radiance or adornment, its beauty and its order.

The Cosmic-etheric in relation to the physical

The elements and ethers

The *effects* of the etheric become phenomenon in the sense world — in its purest form, the blue of the sky — but the etheric cannot be perceived in itself. The concept refers to the holistic manifestation of spirit, whose all encompassing effect we perceive in soul — for example, in a landscape or a winter branch — with our artistic-aesthetic sense, as if coming to us from a spiritual periphery (cosmos). The examples serve to show not only that such percepts always contain something concrete but also the way in which a gesture points to a spiritual direction, from which what is supersensible speaks to us.

Even the earthly or physical is not perceptible as such. It is the *way it manifests itself* that distinguishes it from the etheric. Likewise, an object that, by means of perception, we justifiably assign to the spatial world outside us, always speaks to us concretely through specific sense qualities, and in relation to a place in space in which we move with our physical bodies.

Perceptions of both the physical and the etheric can be differentiated further according to their respective ways of manifesting. In relation to the physical, we speak of the elements, that is:

— earth, the solid element;
— water, the fluid element;
— air, the element that is transparent to phenomena;
— fire, the element that permeates everything.

Likewise, we can consider four different contexts (that is, the etheric):

— through the senses in the present, the totality of all phenomena (light ether), context of appearances;
— through what we perceive as a melody, or what we can find in a chemical process for understanding the inner natures of the chemical elements involved, the totality amongst all processes of transformation (sound or chemical ether);
— through what is active in human or landscape biography (life ether);
— and finally, through the way in which a formative impulse takes effect (warmth ether).

Effect of the cosmic-etheric and the cosmic-astral world in contrast to the effects of earthly matter

Speaking of an etheric world in the terms used in these pages, we refer to the influences that take effect from the world periphery in the direction of the earth. Speaking of an 'astral world', however, we progress, in accord with what the inspired conscious mind observes, from influences coming from the world periphery to specific spirit entities which are revealed in those influences, just as the physical substances of the earth reveal their nature in the forces emanating from the earth. We speak of distinct spirit entities acting from the far distances of the universe just as we speak of stars and constellations when we use the senses to look at the night sky. Hence the term 'astral world.'

From: Rudolf Steiner and Ita Wegman, *Grundlegendes für eine Erweiterung der Heilkunst nach geisteswissenschaftlichen Erkenntnissen* (GA 27), Dornach, 5th edition 1977, p. 7; trans: *Extending Practical Medicine* (London: Rudolf Steiner Press, 1996) Ch. 1, p. 8.

Contemplating all these perspectives forms in us a sense for the invisible organ of the earth that is only experiencable in soul and recognizable according to its orientation in the soul-spiritual surroundings of plants. The effects of this spiritual organ show themselves to us in the way a plant manifests itself, in that they have entered into physical substance.

In what follows we will observe as exactly as possible the ways in which the preparation plants take their earthly shape and how they become a picture of their atmospheric and cosmic surroundings at the same time. By constructing such pictures in the appropriate way with our own observations, we should like to draw into consciousness all that is necessary for the life of the preparation plants when they appear in the environment of their natural habitat, as illustrated above. Constructing the pictures in meditative contemplation can make us aware of a particular aspect of our own being. It is a necessary complement to the reality of what is being observed.

Yarrow (Achillea millefolium)

What must have preceded the life of a yarrow plant and what is presupposed in its appearance when it manifests itself at the peak of its development, as illustrated earlier?

Initially, a spreading of the leaf surface takes place in the first leaf sequence of yarrow, merely as a suggestion, but then it contracts again very quickly. The impression of its having thousands of leaves (*millefolium!*) is produced by the mobile, finely divided pinnate leaflets. These are organized with their tips into a simple overall form of the whole leaf.

At the early stage of growth the leaves are still in a swollen, fleshy state that, however, soon hardens.

It is typical that the individual, finely-structured leaflets of the first order bend so that their upper sides face towards the leaf tips. A leaf-organ arises that is open to the outside, but remains strictly within an overall spatial outline, instead of forming a simple upward-facing, uniform leaf surface. Light and air can pass through it unhindered. Thus no enclosed space forms below a yarrow rosette.

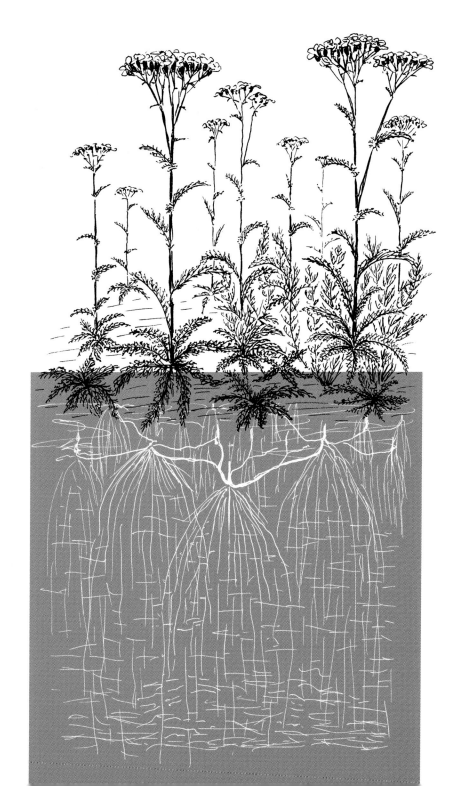

Pinnate leaflets of a yarrow leaf
(left side) from base to tip

Yarrow leaf sequence (selection)

Layered alignment of yarrow
leaflets (schematic)

Yarrow stays on the ground as a rosette in the first year after germination. Towards winter, smaller and smaller leaves appear and they overwinter in the green state. Spring growth begins very slowly. Around the end of May the flower shoot grows upwards.

As with general leaf metamorphosis described above, in that of yarrow the centre of gravity of the leaf shape moves at first outwards and thereafter, on the growing shoot, moves further and further towards the starting point. The leaf base, with its pinnate outgrowths, envelops the stem. Close to the flower, the narrow-linear leaves, often very finely serrated, contract into a simple point.

Flowering starts in July and continues into late autumn. From the centre of the rosette the angular flower stem carries aloft a scaly bud covered with fine silver hairs. The stem divides into flower heads at the top and forms an inflorescence with intricate, branching flower stalks. In doing so, the growth of the secondary stalks does not overtake that of the main stem, yet it does not hold back either. Instead, all the stalks, each with a floret head, extend to a uniform level where they seem almost suspended from above. In this coherence we can observe something similar to the leaf form in that the pinnation is gathered into an enclosed shape.

The inflorescence looks like an umbel, but it does not form rays as a superordinate composite flower, as in the case of a true umbel. Thus the growth of the main stalk does not end in a point out of which equal flower-bearing or, once again, umbel-bearing branchings emerge, but it branches freely, intricately and only coheres in a plane with the heads of the florets. In detail their shape is irregular. But the small knapweed-like (*Centaurea*) compound flowers without pointed tips form, in a so-called false umbel, an upwards-pointing, disc-like, dense inflorescence — a totality of a higher order. Each of the compound flowers comprise tubiform, disc florets and ligulate, ray florets grouped so as to appear as a whole like a five-rayed flower.

The ligulate florets stretch up out of the angular flower bud, extend sideways and stay in this position night and day in all weathers. From their midst dome out the five-tipped tubiform florets to form a small cushion. The scaly envelope of the receptacle always stays somewhat closed. It envelops the flower base which, unlike chamomile, does not bulge upwards.

The white yarrow flowers shine against the dark surroundings, especially in the evening twilight. Sometimes less shining groups are found amongst them. During the day these look reddish.

Scent and aroma are not divulged until the plant is injured. If the flower heads are rubbed, a sharp scent is noticeable. Yarrow tastes aromatic, permeated with a slight bitterness. The same smell and taste qualities, that blend the flower aspect with that of solid earth, permeate the whole plant. It intensifies only in the flowers. The coherence that is a feature of the yarrow shape is repeated in the formation of its aroma.

Seed formation is nourished by the upper bracts and the cone-like green organs of the receptacle that envelops the composite flowers. Finally, these open only a little and the small, completely pappusless, single-seeded fruit (achene) do not easily fall out. A characteristic of yarrow is that the parent plant does not totally spend itself in fruit formation like chamomile. The tough, vertical flower stem does lignify and die, but rosettes formed from tough, vigorous runners remain close to the ground.

Apart from the vertical shoots, there are often branched, horizontal runners. They arise from the lower shoot nodes just below the soil surface, extend through the soil and form further rosettes at their upturning ends. Thus they enable a constant renewal of the growth of the plant, held back close to the ground. If yarrow is repeatedly mown, it can grow into a whole lawn without ever flowering.

Another characteristic feature of yarrow is that it dissolves in the soil into a community of adventitious roots originating from shoots. These form a finely

Development of yarrow flower

branched rhizosphere in which, on one hand, through branching they explore more closely the immediate soil environment and, on the other, penetrate deeply into the mineral subsoil with thin filaments.

If we view all this in conjunction with the flowering of yarrow, we get a stronger impression of its unique gesture.

German Chamomile (Matricaria recutita, Chamomilla recutita)

What processes are a precondition of the picture of the phenomenon of flowering chamomile in May and June?

The very first leaves of chamomile sprout in pairs. In further development, leaves are only formed one at a time. The primary leaves go out immediately as several slightly succulent secondary pinnate leaves with filamentous points. These immediately branch into outwardly stretched leaflets up to the second order. The form gives the impression of openness, softness, raying. The petiole is only distinct at the beginning. Soon leaflets extend from below to the sides. The blades, that still remain delicate but have pinnation of the fourth order, extend

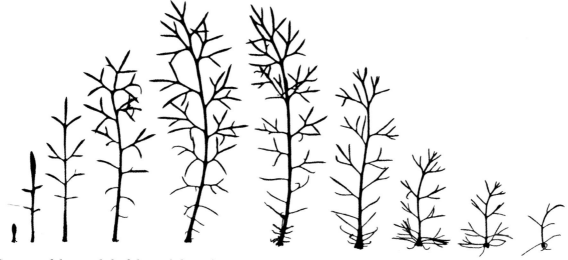

Sequence of chamomile leaf shapes (selection)

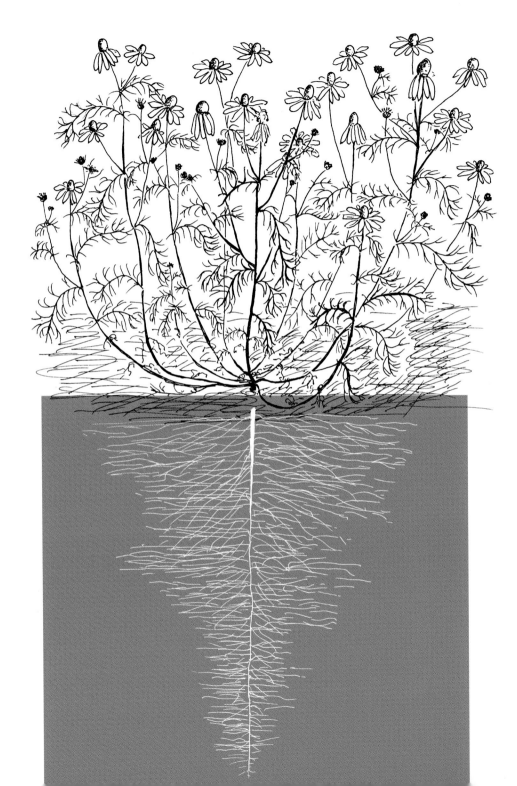

like antlers into space. The pinnation pushes more towards the tips. In the contraction phase towards flowering, the pinnation increasingly extends backwards to the leaf base and wreathes it. The leaves surrounding the stem close to the flower are simpler and may be contracted into fine points.

It is also a characteristic of chamomile leaves that their spreading involves a kind of swelling into filamentous, pinnate divisions which aims, not at broadening the leaf surface, but at dissolution of it. The leaves are softer, moister and fleshier than those of yarrow.

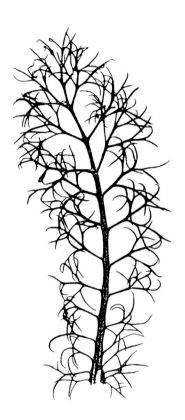

After germination, chamomile roots grow down vertically into the depths of the ground, at first without branching. This shows most clearly after germination in the autumn. Then in spring they sprout several finer, secondary roots equal in size to one another, whereas the upper part of the main root thickens like a tuber. In other words, first comes deep growth then branching, spreading. In yarrow the order is reversed. The part of chamomile above ground begins to develop at the same time as the new root growth in spring.

Chamomile germinates in the autumn and also at this time there is frequently a development of side shoots in the rosette. (Transplanting a rosette in autumn enhances side shoot formation and a richer shooting in spring.) After overwintering, the rosette first extends itself afresh. It becomes bushy, and lush growth emerges. At the end of April the lush shoots, at this stage mostly green, stretch upwards. Initially the stems stand somewhat stiffly erect, then they spread with many branches and form an open spherical shape. Despite the delicate leaves, the greenery at this time seems dense, bulky.

Thus chamomile develops its most luxuriant green at a time when it is growing taller and refraining from flowering. It will do this all the more vigorously the longer it can develop beforehand in the rosette close to the ground and form a long deeply penetrating root. During the ascent of the shoot, the rapidly growing leaves continue in the direction they have already set out on at the beginning of development, forming fine, pinnate divisions. This is quickly followed by flowering. The root remains active only up to the time of flower formation.

When the flowers form, the shoot tip broadens and forms a wide flower base. In the round, spherical bud, the composite flower is enveloped by lots of small,

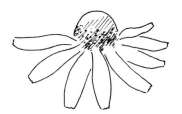 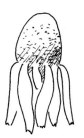  

Development of chamomile flower

green scales. When this opens, the pure white marginal ligulate florets stretch out in a circle like a crown, bordering the yellow tubiform disc florets in the centre. The marginal ray florets continue to grow during the development of the flowers in such a way that they first spread horizontally and, towards evening, bend outwards and down. At daybreak they rise again. This downfolding and straightening up again of the marginal florets is repeated in the course of several days. Such a diurnal rhythmic growth movement in chamomile flowers makes them seem the complete opposite of the fixed, immobile flowers of yarrow.

The inside of the composite flowers is densely packed with a multitude of shining, yellow tubiform florets. While it is flowering, the centre of the flower base bends upwards into a cone. This gives rise to an inner hollow space that is characteristic of the chamomile flower. The centre domes really high until all the tubiform florets have opened. Then the white ligulate floret petals remain turned outwards and downwards until fading occurs. The countless tiny seeds first of all ripen in the outer fringes of the conical, multiple fruit, and scatter like ash as they fall to the ground. Germination can already have begun on the soil surface under the parent plant when the later flowers are only just opening.

Chronologically speaking the main stem flowers first. Then the secondaries grow, starting from the bottom, and overtake (tower above) the central flowers. They begin flowering with their terminal flowers, which in their turn are towered over by their subsequent side shoots. Flowering gradually spreads freely over the whole bushy plant. An inversion similar to that of the opening of the flower head out of the bud is observable in the development of the whole plant.

Thus, a picture first of all arises of life taking shape in airy space in a way that does not lead to hardening, but metamorphoses totally through a kind of inversion into flowering and fruiting. This gesture is associated with a dramatic transformation of taste and aroma. Initially, chamomile tastes mild, like greens. During the luxuriant development of its greenery it acquires a sharp taste that during flowering suddenly changes to an aromatic, typically chamomile scent.

With chamomile's continued, tireless flowering, the plant as a whole becomes exhausted. While it is still in full flower at the top, it begins to fade from the bottom, continuing to put all its effort into its flowers, its warming scent and finally its seed formation.

In contrast to yarrow, chamomile has no capacity for vegetative propagation, least of all underground. Chamomile is the shortest living of the plants dealt with here. Nothing of its life processes is retained in connection with the earth.

*Stinging Nettle (*Urtica dioica*)*

What processes underlie the expression of stinging nettle in its habitat?

Even the cotyledons of nettle are hairy and have stinging hairs. The subsequent leaves form crosswise and opposite, while the square shoot extends. No rosette forms. The leaves are thickly covered with bristly stinging hairs. The basic pattern of their blades is a drop-shape with a sharply serrate border, distinctly separate petiole and steeply descending tips of secondary leaflets. In the leaf sequence, the initially round blade extends backwards towards the shoot from where it joins the petiole, broadens increasingly in sweeping curves to both sides and finally stretches forwards into a lanceolate tip. This movement is repeated in the shape of the regularly out-turned marginal teeth. A typical nettle leaf gives the impression of having started young with prominent venation and having aged rapidly.

Nettle flowers are not very conspicuous. Filamentous, alternately branched flower stalks extend in pairs from the leaf axils. They first lengthen, arch gracefully upwards and outwards horizontally, and finally hang downwards. The tiny flowers scattered on the filaments have no petals and are also thickly covered in stinging hairs.

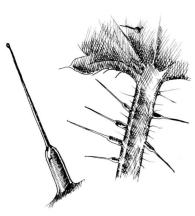

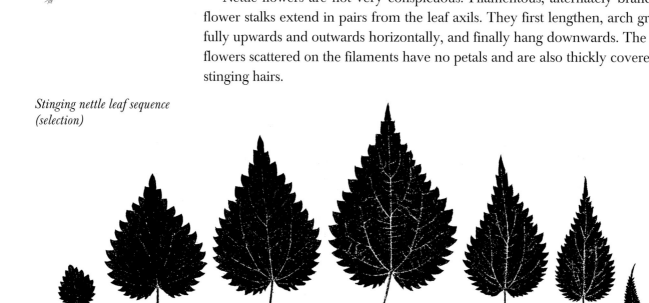

Stinging nettle leaf sequence (selection)

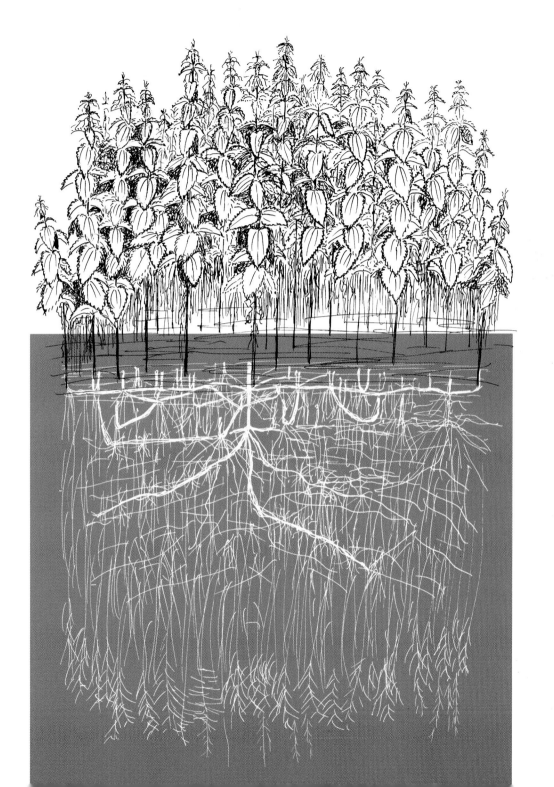

Stinging nettle is dioecious; the yellowish-green pollen-bearing flowers are not on the same plant as the whitish-green fruit-forming flowers, although there are sometimes shoots that bear both types of inflorescence distributed at different levels. The pollen panicles bow out further than those that bear fruit, and are thinner, more stretched and less branched. They disappear after pollination. The fruiting panicles hang down close to the stems and are thicker and more branched.

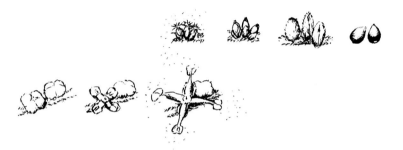

Stinging nettle flowers

Under the magnifying glass, the pollen flowers look like square packets tied up. Four stamens bend away from the centre and are held firm. During growth they spread out further and on a sunny day, when the pollen is ripe, are discharged. They quickly empty their anthers and clouds of pollen blow into the sunny air. Thus, unlike with grasses, the pollen is not passively left to wind, light and sunshine but actively expelled.

The style with its stigma shows as a white hair protruding from the roundish, fruit-producing flowers. In each flower a single, small, light-green, egg-shaped, elongated nut forms.

Nettle occurs where rotting organic detritus gathers, as happens naturally on water meadows. The soil, especially where decomposition processes are at work, is permeated by older, vigorous and very adaptable sulphur-yellow, tough root strands. It is not easy to tell whether they come from roots or rhizomes. From these emerge many, fine, white rootlets. They bring about a well-structured and dried out soil.

In the soil rhizomes radiate horizontally in all directions. These are divided by nodes from which clumps of roots emerge and pairs of new plantlets grow upwards. The resulting network of roots and rhizomes penetrates, for example, the flood detritus of a water meadow, changes it into humus and combines it with the mineral soil. In these kinds of decomposition and transformation processes, nettles create order and have a drying out effect to the extent that, where they grow, fertile soil results. The geometric structure of nettles above ground presents a picture of the forces of order that they develop below ground. They guide the unrestrained processes of decomposition into a humus fit for life.

Nettle grows in dark-green, dense, impenetrable colonies of up to more than a metre high that cover everything below. They form a shady inner space on the spot where they grow like a sheath in the midst of which other plants can barely develop. We all know what happens when we touch nettle with bare skin. Shoot, leaves and panicles are covered with stinging hairs like minute hypodermic needles. Their siliceous tips easily break off when touched, releasing under pressure a fluid that causes intense stinging. The poison brings about a kind of inflammation like the sting of an insect. We become actively involved with and very alert to these plants.

Colours and scents, that could draw attention to the flowering, hardly appear. In touching the greenery a somewhat sourish, fresh smell, slightly sharp with a vegetable tartness, wafts into the surroundings. Against the general greenness of the foliage, the shoot and rhizome look tinged with red and violet. The pollen and older roots are yellow. It is remarkable that here in the rhizosphere, where there is usually no colour, a somewhat flower-like element manifests. This phenomenon is accompanied too by the pleasant smell of the soil transformed by nettle.

Nettle cut before flowering quickly wilts. It shows how the inner sap pressure has kept it upright during growth. The turgor arises from the damming up of fluids. If nettle is harvested later when flowering is already fully in progress, the stems remain stiff. In the intervening period they have lignified. Before the introduction of cotton, nettle fibres were used in fabric weaving.

If a barrel is filled with nettles and water it quickly begins to putrefy and stink. (Nettle juice is used as a fertilizer and a preventive measure against aphids.) Clearly the substances inside the nettle plant are held in a rather labile state in a protein process, and are less inclined to ripening like fruit. For this reason they quickly decompose at death. Witness to this is the smell of ammonia and hydrogen sulphide which is produced.

To summarize, we can say that a characteristic feature of stinging nettle is, towards its base, the interaction of the root-shoot system with what has not yet become soil. Towards the top, this becomes the stiffly upright form of the shoot and the sharply serrate shapes of the leaves and their distinctly rhythmic arrangement. This gesture appears once again, in modified form, in its special way of flowering. Held back in the leaf zone, the flowers express themselves as physically active, blowing their pollen into the air around. However, contrary to other herbaceous plants, it does not speak to our souls with shining and sweet smelling flower organs. Nettle mediates between the material element below and above ground. It transforms the plant material that has fallen down from above into fertile humus and during growth transforms and refines its own substance into the strictly ordered structures of the leaves and leaf positions.

*Oak (*Quercus robur*)*

An oak can take centuries to develop. In contrast to chamomile, which shoots as quickly as possible away from the earth into seed without particularly hardening its form, the oak, precisely because it is a perennial and a tree, roots more firmly in the ground each year and with root, wood and bark forms something durable.

The germination of its acorns takes place underground, and the thick cotyledons with their concentration of fruit-like substances provide all the nutrient of the young plant. In its first year, the seedling acquires only a few leaves on a short, rapidly lignifying shoot. Initially, the more vigorous and quick-growing

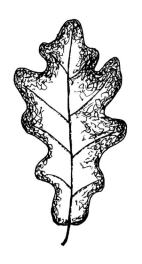

root also begins to lignify. Flowering and fruiting only happen after many years. Oak begins to flower only when the tree and its crown have attained the complete form.

Its asymmetrical, round-lobed, leathery leaves develop late in the spring and in winter some of its dried-out leaves frequently remain on the tree. The first leaves on a new oak shoot are only slightly lobate. In later leaves the surface is more divided, but without the petiole extending. In pedunculate oak (*Quercus robur*) the leaf base is bilobate, cordate, auriculate. The lobes become bigger towards the tip. The foliage grows predominantly in clumps at the end of the twigs and encloses the inner space of the tree like a roof. Thus, together with the gnarled growth of the branches this makes the crown of the oak irregular, free and open to the surroundings. It encloses a space that is nevertheless permeable to light and rain. Therefore, other plants, for example hornbeam, many woodland shrubs and herbaceous plants, can thrive *under* oak.

The inconspicuous, small, green inflorescences often appear unnoticed at about the time that dandelion flowers, along with shooting and the beginning of growth of the stems and leaves. The pollen-forming inflorescences hang from the axils of lower leaves of young shoots, whereas those that form the seeds sprout from the axils of the leaves close to the tips. The pollen flowers are grouped as a catkin round a delicate shoot that stretches into space and hangs down, flowering loosely with yellowish-green perianths. Each contains six stamens with short filaments. The equally small female flowers are attached singly, doubly or in groups of up to five on bare stalks. Their tongue-shaped stigmas are often red. The flower expresses no striking coloration and emits no scent.

The ovaries are enveloped by a sheath, the cupule. The acorn that arises from it forces it open and is subsequently seated in a hemispherical bowl shape. By the end of September, it is fully grown and in October it falls off. The winter buds are gleaming brown, broad, somewhat angularly egg-shaped and bare. The lighter or darker brown autumn foliage often only falls off during the winter.

A young oak is not yet gnarled. With its smooth bark and straight shoots, often arranged radially in a whorl shape, it sometimes almost resembles a cherry tree. According to location, the attribute of a central trunk, characteristic of the young tree, disappears. In the development of the tree the central stem does not simply continue to grow but instead forms several main branches. Inhibition at the twig shoot tips is common. This is then continued as irregular lateral branching. Thus the direction of growth of the branches changes repeatedly and they become increasingly angular. They are also rigid and brittle and not elastic like those of ash. (They give the impression that they are more inclined to snap than bend.)

The branches sometimes die as a result of shading caused by the tree itself as it gets bigger. Mighty trunks of several-hundred-year-old oaks are sometimes rotten within (fat stag-beetles etc. work through the rotten wood). The huge, thick old branches often lie around broken and in splinters. It all seems dead, and yet new, vigorous branches project from all over the trunk. The familiar gnarledness of oak's overall form reflects not so much a growth pattern as an interaction between new shooting and growth and rigidifying and dying.

Root formation in oak has a mind of its own too. The tree drives a vigorous tap root as far as possible into the ground. Then, starting at the top of it, lateral main roots grow downwards at a slant. At a later age the roots grow horizontally in all directions and send sinker roots down vertically into the subsoil, reaching deeper than the roots of most other trees.

Young twigs are somewhat angular at first, hairy, shining olive-brown, and later form smooth, shining, green to whitish-grey reflective bark. This later gradually becomes grey-brown to blackish, forms light-coloured tears lengthways and is transformed into thick, deeply-creviced bark. In this pronounced, cumulative mineralization process organic substance is expelled from the living realm and dies off. The ageing process of oak is particularly impressive in the bark.

The tart, tannin taste of the reflective bark gives an immediate and vivid impression of the contracting tendency of substance formation in an unspecific hardening and consolidation. Later in bark formation, this process stops, the corresponding taste disappears and, as the bark becomes irregular, its lime content is largely amorphous, but is deposited in microcrystalline form in measurable quantities. The wood of oak, with its relatively high resistance to weathering influences, is rich in minerals.

For oaks that are suitable for the preparation, see Appendix.

Something similar happens in fruit formation only in the reverse direction. In contrast to the bark substances that are deposited outwards, fruit substances are stored within. Although they do this close to the seed germ, they nevertheless remain outside it. They are alien to life processes, yet in a condition in which they are easily taken up again by them.

Oak's relationship with animals is especially interesting. The compact acorns form not only a new plant, but also provide many animals with valuable food, especially in the autumn and winter. Oak also has a diversity of relationships with insects, particularly through the secretion of its sweet sap and with its tendency to form various kinds of gall. Gall formation starts with an insect spraying a poison at a particular place on a growing organ (leaf or bud). The resulting growth inhibition is diverted into the growth of a specific fruit-like structure, that serves the development of the insect's larva. In this way, gall-wasp eggs turn into an ovule alien to the host organism, taking the form of the 'fruit' formed by the oak. This supports the insect's development and maturation. This is a further aspect of the aforementioned way that oak takes shape; one in which dying is overcome by new formative potential. Around oak there is rustling, scurrying and hopping. The tree gives protection and nourishment to a multitude of animals.

It is in conformity with the gesture of oak's life processes that the branches are thick; the seeds are large and heavy and plump to the ground. Powerful, rampant tree-forming forces rising from the ground are active in the irregular, open growth-pattern and in the composition of the branches and trunk formation. The bulky materiality of the thick branches represents an extension of the trunk. This force is attenuated outwardly in the bark and inwardly in the wood, and diverted into purely physical, mineral tendencies. The dying and hardening awakens new growth processes in the reverse direction. Oak retains the power to send out new growth right into old age. Of all our broad-leaved trees, it lives the longest.

To summarize, we can therefore state that, in life and overall form, oak expresses the way in which within it constant ageing and dying to the point of mineral deposition set free and enhance a capacity for new burgeoning. Both dying and new growth originate in the cambium.

*Dandelion (*Taraxacum officinalis*)*

A long development precedes the impressive atmosphere of the shining, yellow flowering of dandelion as indicated in the previous illustration of the lushly growing meadow. If a dandelion germinates in autumn, it cannot then flower in the following spring. It first needs a summer in which it orients itself to the sunlit surroundings with its continuously spreading green leaves, and in doing so acquires the capacity to penetrate deeply into the subsoil with its root. Towards autumn, what has been brought about over the summer is developed further. Thus, in the crown of the rosette that is kept tight against the ground, spherical flower buds form surrounded by both a dense, white felt of hairs and the extended bases of the many smaller leaves that formed towards winter. This rosette centre, with its new flower buds, is drawn into the ground by the root that

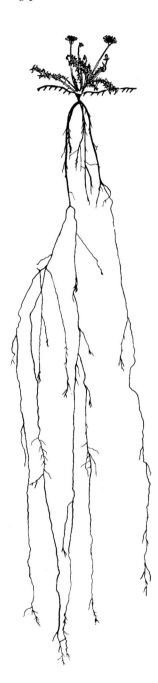

in the meantime has enlarged to a latex-rich tuber. Thus the suppressed main shoot, from which the wealth of leaves (a hundred or so at that stage) emerged, becomes very like the root. This 'inhalation' of the whole plant, in the rhythm of its part of the earth towards winter, is necessary for the rapid appearance of the dandelion in the spring.

The whole process is reflected in its leaf sequence, which starts with elongated rounded cotyledons followed by primary leaves with petioles that constantly transform into thin, rounded blades. While the blades in turn spread further, the spreading also noticeably includes the petiole and forms a leaf seam right to the base. Lengthening of the petiole and spreading of the blade surface here blend together.

Differentiation begins with the narrow teeth on the leaf margin. Later in the sequence these become very jagged points with their tips towards the centre of the rosette. They arch outwards and are frequently dentate again towards the tip. This gives rise to the characteristic sharp teeth of dandelion (cf. the common French name *dent-de-lion*). The centre of gravity of the leaf surface remains in the upper region, even when the bare, soft leaf shows pronounced pointing later in the sequence. In the second half of the year, the leaves become narrower, flimsier, the individual teeth seeming more independent and the blades extend equally on each side of the base.

From the end of April onwards the soft, latex-carrying tubular stems grow upwards from the inner leaf axils together with the new leaves, that is, the stems emerge from the heart of the rosette where the buds overwinter. This process reflects the jump in the leaf sequence, which shows no intermediate forms between the biggest rosette leaves and the sepal-like bracts of the enveloping receptacle. The buds form at the ends of the hollow, sappy stems as small, green, round heads surrounded by several layers of scale-like, linear bracts or sepal-like leaves. These open in sequence and bend downwards. Above them the individual flowers of the large compound flower stretch out radially as shining yellow, long, narrow tongues. Each flower head has between a hundred and two hundred florets that gradually open in sequence from the margin inwards. The flowering of the whole plant and a patch of dandelions happens almost

Dandelion leaf sequence (selection)

simultaneously in a period of one or two weeks. Whole meadows and roadsides are covered with radiant flowers appearing as if in a single plane against the fresh green of the background. They are visited not only by bees but also butterflies, beetles and flies.

The inflorescence of Compositae is closely held together as if in an undeveloped state. The substrate for the large colonies of florets in the compound flower forms a flower base that extends far beyond the diameter of the stalk. The golden-yellow compound flower opens as the sun rises in the sky, turns during the course of the day towards the sunlight and closes in the afternoon still facing upwards. In periods of good weather, each flower head opens and closes over a period of several days. In doing so, the individual flowers of the composite unfold from the margin towards the centre. If it is a dull or rainy day, the flowers stay shut. Thus the flower heads show a sense-like sensitivity to light.

During flowering and fruiting the flower stalks grow longer and lift the flower heads surprisingly high. This is not always noticeable because the grass and other plants beside them grow taller too.

When finally the innermost florets of a head have faded, it no longer opens. The ligulate florets wilt, go brown and dry, and are pushed out of the receptacle that is closed like a bud.

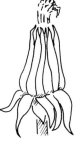

Development of dandelion flower

The scaly bracts of the enveloping receptacle remain fresh and take care of the ripening of the fruit. Thus for several days it seems as if nothing is happening until one morning the bracts open out completely and fold back towards the ground. The flower base arches upwards to form the base of the multiple fruit. The base everts with the fruit to form the familiar dandelion clock. The result, right to the very last detail, is a transparent, filigree sphere, as if rigidified like a crystal, with the seeds tightly packed together in the middle. Now we realize what it has been preparing within. The hair-thin calyx of each individual floret has grown into a tiny parasol (pappus) situated on the tip of a very thin stalk that extends out of the top of the fruit. The clock is an image of an orientation to the entire periphery. It relates to wind, like flowers relate to light. The brightly shining pappi carry single-seeded (achene), dark fruit far into the surroundings. Thus the growth of dandelion continues not only during flowering but also during fruit ripening. The seed-like fruit germinate in favourable lighting within one or two weeks.

As early as during flowering, new rosettes grow from the outer leaf axils of the rosette on the ground and, as with immediately after germination, start off with simple leaf forms. Thus, in May dandelion leaves offer a bewildering array of forms. The old stems wilt and disappear shortly thereafter.

Dandelion rooting is dominated by the penetration of its main root straight down into the soil, but at a certain level this branches and penetrates further downwards. The root passes through the richly humous soil into the subsoil and only there begins to spread properly.

The polarity in dandelion's capacity for regeneration is worthy of note. Not only can new plants regenerate from the cut root, but also the fruit can form when no cross-fertilization has taken place. Thus dandelion is always complete, with no gaps showing.

The taste of the root and the full grown leaves is somewhat bitter, but the young leaves are palatable and indeed very appetizing as an addition to spring salads. In contrast to the flowers of its close relative chicory, the flowers do not taste bitter but rather sweet. The fruit tastes nutty.

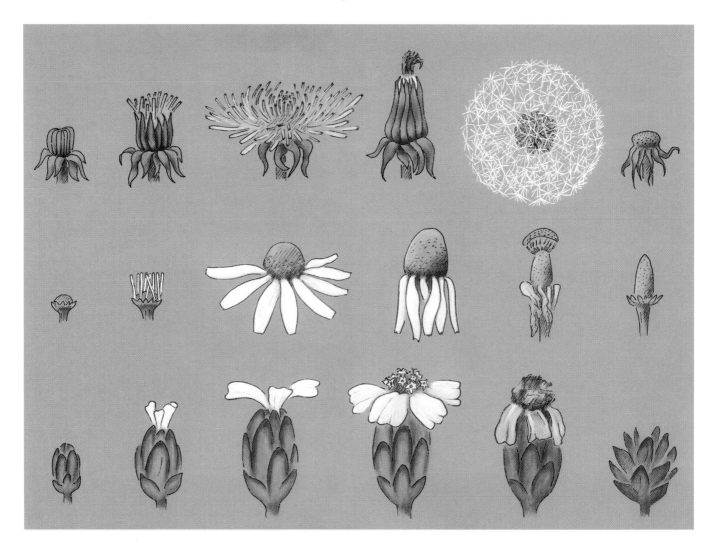

Flower gestures in three Compositae

The compound flower head (capitulum) is surrounded by small green leaves (bracts) which form the calyx-like receptacle.

In dandelion these are straight at first. Then they bend downwards — the outer prior to flowering and the inner during it — and remain mobile. With chamomile they extend sideways and remain like a crown. The receptacle of yarrow stays shut like a pine cone.

Dandelion's long, yellow, ligulate florets emerge one after another and remain mobile. In chamomile only the marginal ray florets move, whereas those of yarrow stretch out of the tip of the bud-like receptacle, spread sideways and remain until the fruit ripens.

Dandelion flowers open and close over several days in a row. By contrast, the diurnal movements of chamomile flowers comprises the marginal ray florets folding outwards and downwards. After dandelion has flowered, the inner part of the receptacle closes and reopens only after the fruit ripens. During this the flower base everts to form the intricately differentiated sphere of the 'clock' with seeds in the middle. The hair-thin calyx of each floret grows in concealment into a fine stalk. The brightly shining tiny parachute carries the fruit far into the surroundings.

The base of chamomile flower arches up with the golden-yellow shining tubiform disc florets to form a hollow space. Its minute seeds ripen from the outside inwards and spray down onto the ground. The scaly receptacle of each yarrow flower only opens a little though not until the fruit has ripened through drying.

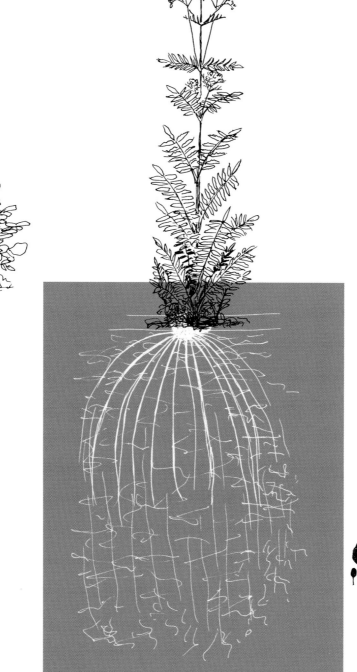

1st year

Common Valerian (Valeriana officinalis)

The sweet, musty wafts of the scent of common valerian flowers needs to be experienced in its natural surroundings along with the picture of it. What it gives off in refined form there depends on what has happened previously during the course of the year.

After germination, the first thing to emerge is a long-petioled leaf with irregularly rounded blades, followed by lush, soft, intricate pinnate forms. In sequence, the leaflets become longer, more pointed and frequently irregularly dentate, whereas the petioles continue to lengthen. Towards autumn the petioles shorten again, the indentation dwindles, but the differentiation of the pinnate leaflets does not recede. Thus an open rosette forms with a spiral arrangement of leaves that disappear again over winter.

During this period the seedling forms a small tap root that soon dies. The shoot retained in the rosette expands gradually into a rhizome. Out of that grow numerous round adventitious roots of equal thickness from the top down,

2nd year

Valerian leaf sequence over two years (selection)

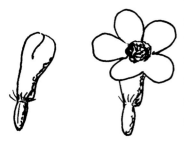

which penetrate the ground in an arch formation. Towards winter they show more clearly how they enclose a spherical space in the soil. In both the skin of the rhizome and in the somewhat tubular, enlarged roots is concentrated a musty, aromatic smelling sap with a camphor-like, invigorating taste. The smell is similar to the scent of the flowers, albeit not so sweet and rather mustier. It is related to smells that arise during the ripening of many kinds of fruit. The rhizome forms secondary shoots either straight to the side or below ground, and sometimes sends out runners above ground.

In the second year the development stages of the rosette are first repeated. Vigorous, juicy greenery spreads out. Now the leaves are opposite and the leaf shape is different. There is a greater number of imparipinnate leaflets which are now longer, narrow and less dentate. The dentation increases in the leaf sequence and becomes thinner, sharper, but it can disappear altogether in some varieties. Before midsummer, the flower shoot stretches upwards. The relatively few leaf pairs, that arise well-separated from the long, forked, hollow stem decrease in size towards the top and become more pointed and compact. They finally disappear in the transition to inflorescence.

The inflorescence is first visible as a whole at the end of the shoot in the form of a large, undifferentiated bud which is still enclosed by the uppermost foliage leaves. But soon it becomes clear that this is an assemblage of many individual buds and that an umbel-like panicle will form. The individual flowers at the rounded bud stage are fully enveloped by three small, delicate leaflets and tightly packed together. Their colour is delicate pink to reddish. The opened flower head reveals five white to bright pink coloured sepals joined at the bases. Whereas the vegetative organs of the plant do not at first have the typical valerian scent, this now reappears in refined form at the opposite pole to the root. The flowers do not all open at the same time. During the first flowering, fading and fruiting, the inflorescence divides further, and at each stage forms new flowers. Thus fading takes place before the greatest unfolding of the inflorescence, whereas the flower stem forks again in stages at each flower. This can gradually give rise to up to two thousand individual flowers at higher and higher levels in the panicle. Growth of the inflorescence continues during flowering and fruiting, opening it up through the continuous forked branching of the decussate flower stalks. The panicle thus gradually forms a delicate vault that rounds off valerian at the top. The seed head then appears open and airy.

When the vault of flowers fades, dark, claw-like structures curved towards the centre appear on the upper edge of the ripening fruit, initially rolled up, forming a kind of wreath. These are sepals which, during ripening, extend outwards and form downy, airily light, pinnate hairs. By comparison with the parachute-like pappi of dandelion, these are less strictly formed and are not positioned on the end of stalks. Thus the fruit head becomes a delicately ordered, drying structure comprising thin, reddish fruiting stems with single-seeded fruit whose feathery pappi gleam in the sun. On a dry day, this down is caught by the wind and blown over the countryside — fruiting into the surroundings. The process is accompanied by the musty valerian smell that is no longer given out by the flowers but by the wilting, vegetative parts of the plant.

Valerian distinctly separates its vegetative and generative phases. First a rosette forms together with the roots. After overwintering, the shoot ascends with its strictly arranged leaves that become increasingly refined in shape the closer it gets to flowering, fruiting and giving off scent. Whereas the vegetative growth has hardened in structured form, and stands out prominently from the surroundings with its linear, vertical stem of up to two metres in height, the saps have been refined into the scent of the flowers. Towards autumn, valerian, a perennial, sends out new rosettes beside the old flower shoot or forms runners from which new rosettes emerge. As one of the earliest plants to start growing in spring, it develops new leaves in the rosettes from the midst of which, after a while, flower shoots stretch upwards.

Valerian has a strong tendency towards rampant growth. This increases in cultivation or when the soil is naturally rich in humus and light. There it begins to grow luxuriantly and depart from its form, first in the basal greenery and later when the shoot grows up. The occurrence of double leaves, two instead of three at a node, the tendency for banding, thickening and other deformities, all increase but with flowering, the opposite tendency to rampant growth becomes more and more active. It reaches its peak, as described, in the filigree-like formation of the inflorescence, the arrangement of the stem leaves and the hardening of the shoot. If valerian grows in damp margins of undergrowth or on water meadows and is there exposed to plenty of sun towards flowering, the two tendencies are held in balance. In shady locations the flowers are predominantly white. In the sunnier places they often turn pink. The stems, and to some extent the leaves, are increasingly tinged with red. This increases towards autumn when, during wilting, the plant smells strongly throughout, becomes rigid and lignified.

Valerian's scent concentrates in the skin of the rhizome and in the root strands. This happens most strongly when the rosette is preparing for autumn and winter. On drying, the smell of the rhizome increases further. Fresh leaves smell like greenery, but on drying, the typical valerian smell develops. Thus the whole plant is permeated by it in latent form. Dried samples of the plant in a herbarium can give off the penetrating smell for decades through paper or plastic film. Cats love the smell of valerian and like to tread in the leaves. The smell has an intoxicating effect on them, rousing their instincts.

The smell changes during the development of the plant. The fresh flowers release a more flowery scent. The pressed-out juice smells more typical of the plant. The valerian smell occurs most strongly at the transition between rampant growth to fading and dying.

In autumn, the smell that is given off by fading valerian, particularly in the morning, is similar to smells that are associated with healthy humification processes or fruit formation. (Interestingly, most of the familiar aromas of fruit, for example banana, strawberry etc., are esters of valeric or isovaleric acid.) These aromas are both an expression of the fruit ripening and of general ripening processes in the soil. Thus the rhizosphere as a whole, already described, is a picture of the special capacity of valerian to stimulate ripening processes in the soil.

Common Horsetail (Equisetum arvense)

What conditions does horsetail need to thrive and what are the features of the habitat that enables it to form such fine silica structures?

Stiff structures (sporophylls), at first reddish brown, reminiscent of parasitic plants or fungi, emerge in early spring out of the bare earth. Their lower part is divided into nodes each separated by a wreath of small, erect, clasping, dark-brown leaf tips. The spindle-shaped spadix-like cones are evenly divided into hexagonal areas. Each of these covers six sporangia sacs. What happens to the spores is largely too small for observation with the naked eye (see box).

Illustrations of the development of common horsetail

Horsetail has no flowers or seeds. The early stages of its visible development take place on a microscopic scale in water. After several days, the fertile (sporophyll) shoots (1) have extended to 20 centimetres and the spores have become ripe, the sporangia sacs (2) split longitudinally and the dusty spores (3) are released. Some are caught by the wind and blown away. If they fall in a wet place they begin to germinate (4). A root-like tip emerges and a greener, more irregularly shaped, alga-like prothallus grows onward (5/6). (At this stage the horsetail is actually a green water-plant.) The prothallus differentiates in two ways. One (5) develops small clumps which dissolve and release lots of circular spermatids. Each of these have two flagellae and move forwards in water. Other prothalli (6) form archegonia with egg cells that are fertilized by the spermatids. After fertilization the young horsetail plant develops and stands vertically (7). Usually sporophylls and vegetative (infertile) shoots regenerate from the underground rhizome.

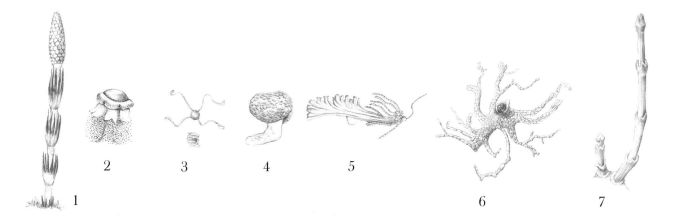

1   2   3   4   5   6   7

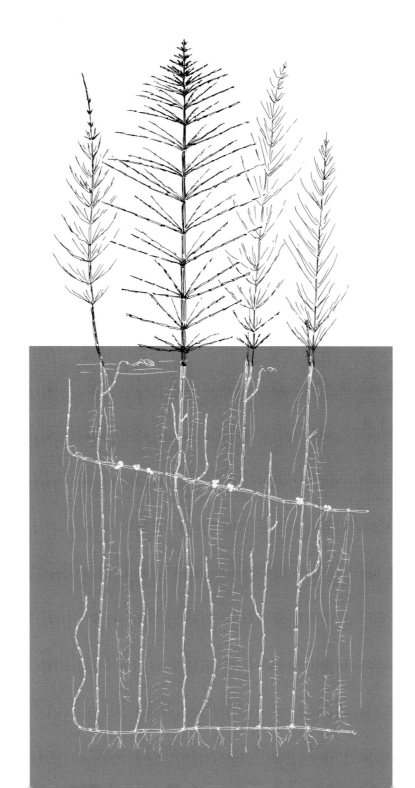

While the spores are ripening, somewhat thinner, sharp, green shoots which have wreathes and side branches from the start appear in the vicinity of the sporophylls. These vegetative, infertile shoots are what we usually recognize as horsetail. They extend upwards almost like telescopes and spread sideways forming radial whorls, divided regularly into several levels, like a pine tree. The individual sections of stem are interlocking. The stem is ribbed and feels rough to the touch. The leaves hardly develop, remaining as tiny brown scales. They encircle the end of each section of the main and secondary shoots like little crowns. The leaves, stem and secondary shoots are leafy green. In this respect, horsetail behaves like a succulent, albeit with greater flows of water and air through it.

In contrast to those of other horsetail species, the fertile and vegetative shoots of common horsetail are spatially and temporally separated from one another.

During further development horsetail plants stand erect, separate themselves from the ground and grow towards the light. The stem and its stem-like secondary shoots replace all important functions of land plants in a way that is modified compared with flowering plants.

After germination as it grows upwards with astonishing speed, the young horsetail plant sends rhizomes channelled with air ducts down vertically into the subsoil. They sometimes reach three or four metres down. When they meet a wet zone (groundwater, for example) they continue to grow horizontally. The formative pattern of the rhizome resembles that of the shoot, although the latter is simpler and not green. The rhizomes are light-coloured at first, turning brown and later dark-brown. From them emerge adventitious roots. There are no primary roots. In fact, unlike flowering plants, there is no proper root system.

Propagation of horsetail is primarily vegetative. Numerous shoots rise vertically in a simple rhizome-like way from the underground rhizomes. As soon as they reach the light they send up long green shoots up to sixty centimetres in length. These can go on appearing right into the summer. From late summer to late autumn the underground rhizomes form tuberous swellings which obviously contain food stores that serve the plant in spring. They are egg-shaped structures in groups of up to four.

Silica structure of horsetail after ashing

Like all plants, horsetail is supported by a framework of carbon compounds, but the fine structure of its surface is made up of silica, partly in epidermal cells and partly deposited on the outside. This does not crystallize out, but solidifies as opaline glass and in places forms lenticular bulges that focus sunlight on to the rows of chlorophyll molecules. On ashing, horsetail leaves behind a fine, white skeleton of silica. The silica is taken up from the ground with the water. This requires large amounts of water to flow through the plant, which are actively discharged at the top. It is a special feature of horsetail that it reaches the richest supply of water in the subsoil and from this forms itself in fine silica structures up in the light.

By comparison with other horsetail species, common horsetail withstands strong sunlight and drier conditions near the soil surface, but it needs an uninterrupted flow of water from waterlogged ground below, especially in the growth phase, but if it is picked and put in a vase of water it quickly wilts. Only common horsetail and giant horsetail (*Equisetum telmateia*) form separate sporophores that do not assimilate. Otherwise the sporophores are on the tips of all, or some, of the green shoots. An example of this is wood horsetail (*Equisetum sylvaticum*).

8. A Comparison of the Gestures of the Preparation Plants

Based on the foregoing descriptions, in this chapter we compare the life rhythms typical of the preparation plants, taken in pairs, paying particular attention to the organs described later. Our aim is to gain a clearer view of the respective gestures of the plants, their qualitative, cosmic orientations, from the point of view of the effects of the preparations and their overall composition.

The life of the plants in the elements as a key to understanding their cosmic orientation

When a plant unfolds in front of us it forms its substance in a particular inter-action of the elements — earth, water, air and warmth. When it starts to grow, its meristem seems malleable, pliable like a droplet, and takes on a particular shape as it increases in size. Earth-water elements become solid matter. These are the elements in which the perceptible plant lives. On the way from the fluid, pliable state to the solid that separates out from it, a clearer and clearer picture of order appears; a certain beauty in the manifold qualities. In this sense, the resulting phenomenon of the developing plant becomes airily transparent to its cosmic orientation. It is the same process that in the air allows us not only to see objects, but also perceive atmos-pheres, that is, landscape moods revealed as a whole. So from the way the plant takes shape, something emanates that speaks specifically to our soul, permeating us in a particular warming or cooling way, and which also com-prises the quality of substance.

The inner nature of the elements supports particular orientations

Since Aristotle's time, people have spoken of four elements, earth, water, air and fire. Nowadays, when people only accept what can be thought in terms of fixed pictures, we refer to aggregate states. In doing so we base immediate phenomena on pictures of the smallest particles or points of force in which, when combined (aggregated), we seek the cause of the qualitative differences. Such a way of thinking is based only on what can be contributed by solid state mechanics. But understanding qualities of elements is already assumed in this mechanical thinking and the so-called explanations that go with it bring nothing new to an understanding of the percepts in which we have a differentiated relationship to the solid, fluid, gaseous and heat states. The elements are not what we usually understand by them. We appreciate their inner nature only in the various ways of encountering the phenomena of the sense world. Mindful of this, we acquire various capacities or ways of approaching the world of appearances in self-observation:

Without being able to take a firm hold of things in pictorial thinking, we would not consciously experience the reality of solid objects. Objects are changing appearances for perception alone.

To understand the fluid element of the world, we need the faculty of making the pictures that arise in the sense world transform into one another in such a way that we recognize in this process something that is essentially always passing through.

We appreciate air as an element that, like the lens of the eye, withdraws in order to leave space for the phenomena and allow us to grasp the atmospheric aspects of them.

We are most closely connected with warmth, which as an element permeates everything. When two things meet, something goes from one to the other. It announces itself as warming or cooling in the warmth that lives in us.

See Jochen Bockemühl (ed.): 'Elemente und Äther-arten als Betrachtungsweisen der Welt' in *Erscheinungsformen des Ätherischen* (Stuttgart: Freies Geistesleben, 1977); trans: 'Elements and Ethers: Modes of Observing the World.' In *Towards a Phenomenology of the Etheric World* (Spring Valley: Anthroposophic Press, 1985) and Jochen Bockemühl, *Ein Leitfaden zur Heilpflanzenerkenntnis*, Dornach: Verlag am Goetheanum, I, 1996; II, 2000; III, 2003 (English translation in preparation).

If we look at plants in this way, it will enable us to grasp what Rudolf Steiner had in mind when he spoke about them in the seventh lecture of his Agriculture Course:

the plant has an immediate relation to earth and water … through a kind of breathing and through something remotely akin to the sense-system [the plant] absorbs into itself directly all that is earth and water … The plant lives directly with the earth and water. … the plant does not consume but, on the contrary, secretes — gives off — the air and warmth … Air and warmth, therefore, do not go in — at least, they do not go in at all far. On the contrary they go out; instead of being consumed by the plant, they are given off, excreted. …

The life of yarrow is constantly connected with the earth, refining and concentrating the fluid, mobile element. Chamomile tends to leave the earth, transforming the ascending substances into scent and warmth. Both yarrow and chamomile have many similarities. For example, as they grow out of the earth both form finely sculptured pinnate leaves, but if we study the way they are formed, their contrasts become clear.

The fine division of *yarrow* leaves, appearing ready-made through its intricate branching, seems fluid-like, but if we cast our eye over the tips of the leaflets, we see a clear image of an outline which appears at first as a closed leaf form. The same repeats in the branching of the inflorescence and the seemingly two-dimensional (corymbose) finely-structured flowers that stand upright and are kept together. These appear, radiating atmosphere, at the peak of the plant's development and continue right into autumn, while the stem is lignifying. The flowers remain unaltered for a long period and the ripening fruit stay held together by the receptacle of the flower head.

The gesture through which the finely divided form arises from the fluid element, and is then retained and fixed in an overall shape, is further intensified and enhanced by the way in which the tartly aromatic taste, that permeates the plant from the outset, is held back in the saps and is not given out as a scent unless the plant is crushed or rubbed. On the whole, yarrow keeps its vitality close to the ground with its runners and rosettes, and permeates the soil with a fine root system reaching deep into the subsoil.

By contrast, the delicately succulent leaves of *chamomile* appear as curved filaments, pinnately branched, flowing outwards instead of held together, and much more prominent than those of yarrow. In the course of its life, chamomile deeply penetrates the ground with its tap root, but then wholly dedicates its vitality to rampant growth. Then it turns itself inside out in flowering and fruiting so to speak, while wilting from below and renouncing its connection with the earth.

What happens to the saps rising from the ground is expressed at its clearest in the enormous transformation of the taste, which in the leaf goes from neutral, like greens, to a sharply leafy taste and ultimately changes further in the transitions to flowering and wilting, ending by being given off as an aromatic scent. The gesture of chamomile is one of substance rising from the earth and being constantly refined into air and warmth.

Stinging nettle is connected with what has not yet become earth and creates a definite rhythmic order in its growth. Valerian supports ripening processes in the ground and balances rampant growth and structure.

In its natural habitat *nettle* appears well separated from other vegetation, its greenery spread out. At the base it creates space in which disorderly piles of decaying substances transform into humus. In its orderly foliage at the top it forms a substance that decomposes easily and generally promotes healthy plant growth. By these activities nettle supports, in a special way, the interaction between processes of the universally fertile humic soil and flourishing vegetation.

Valerian, which sometimes grows at similar sites, does not become involved in humification, but promotes soil maturation processes that are important for the fruit formation of vegetation. The sweetish musty scent of its flowers, that in vegetation is generally associated with the formation of juicy, fleshy fruit, is diverted to the roots. During fading and drying, this smell also permeates the formerly green vegetative organs. In autumn it scents the air of its growing site.

Oak creates a space in which it takes life processes to the point of dying and increases the potential for new growth to begin. Dandelion penetrates the subsoil, feeling its way, and in flowering turns rampant growth towards the sun.

Oak and dandelion immediately strike us as very opposite to one another in form. Their habitats are mutually exclusive. *Oak* forms a physical space in which substance formation tends to the mineral state on one hand and on the other leaves space for vegetative growth. On a large scale, this happens in the formation of its gnarled shape and in creating space for the development of pro-lific undergrowth. On a smaller scale, it appears in the dying and deposition of the tannin-rich and, later, lime-rich bark, and in the constant new shooting, both of which originate in the ripe cambium. This gesture is also consistent with oak's capacity to react to being punctured by gall-forming insects.

Although oak acts in a special way to create space for itself, like other plants, it nevertheless also needs the appropriate landscape atmosphere that enables it to thrive.

Dandelion in contrast grows in open, sunny meadows. Its gesture is evidenced in two ways: firstly by its reaching deep into the mineral subsoil with its tap root, there raying out its root system, and, secondly, by its flowering in relation to the sun, which coincides with the beginning of the lushest growth of meadow vege-tation and transforms itself into refined structures that are permeable to air. Prepared by winter bud formation, in which it concentrates what it has taken in from the surroundings in the previous year, dandelion has a special capacity to combine and balance rampant growth with flowering.

Common horsetail is associated with subsoil water and takes shape in finely-structured silica.

Horsetail holds a special place in this study as it is not one of the plants used for the compost preparations. However, it was mentioned several times in the Agriculture Course in connection with its silica effect in the sense of a healing plant.

What corresponds to flowering in other plants happens with horsetail largely in the water element in the ground. It does not emerge from the ground in swelling, spreading growth. It comes to a stop in a mineral structure that, in con-trast to oak, does not comprise finely crystalline deposits of lime, but a finely structured framework of opaline silica deposits resembling glass. With silica's special property at the level of greening plants, these become airily transparent and thus help the plants to bring their cosmic nature to manifestation.

9. Organism and Organs

An interaction of processes is not yet an organism

People today are comfortable with regarding the whole earth, a landscape or a farm as an organism. This is because they sense that in these they are dealing with something as a whole. But what exactly is this *whole* that we call an organism? What do we know about it that gives us cause for such a view?

We need to be clear about this and consider the implications of it if, with Rudolf Steiner, we are to understand a farm as an organism and as a kind of individuality, and in so doing, try to grasp the meaning of the biodynamic preparations discussed here.

We normally live with the contradiction that, although we realize that cars are not organisms, as soon as we deal with an organism in science, we describe it as a mechanism and use the same ideas of mechanisms, with self-regulating, chemical or mechanical processes, as we are used to doing with cars. But this does not make use of any concepts appropriate to an organism.

The saying, 'a whole is more than the sum of its parts' also applies to cars. But what does this 'more' comprise? Cars are made with an idea for constructing them according to which parts are made and fitted together. But the materials on which this idea is based are not from a common physical germ, as is the case in plants. In fact the materials have little at all to do with the idea. Even the connections of a mechanism with its surroundings have to be inbuilt from outside.

Experiences, usually subconscious, of organism and organs are largely attributable not only to observation of animals and plants but ultimately to ourselves — that is, to the way we are situated in life as human beings. We experience ourselves as a unity and observe how we look out on a diverse world

Fundamental categories of experience for comprehensive pictorial-imaginative cognition

In what we may call 'modern homelessness', or the feeling of isolation among people today, there is a significant opportunity and challenge for us to make use of a capacity of imaginative perception, hidden in all of us, to actually grasp the phenomena that at first are only present physically, and in this way establish our own relationship with the world from within outwards.

At the beginning of his book *Theosophy**, Rudolf Steiner, using an extract from Goethe, presented three fundamental categories of knowledge through which each person can become aware of their place in the world with body, soul and spirit:

We perceive objects that we are looking at with our senses. We discover something present in the world. This kind of knowledge of an object, right through to the finished concept, is mediated by *body*.

What we perceive evokes our sympathy or antipathy and corresponding tendencies to make judgements.

Through this we make the world into our own affair. This comprises *soul*.

The secrets of the activities and being of the objects that we perceive reveal themselves to us. Such knowledge gained appears always as a goal towards which we strive. It is the fundamental knowledge, constantly active in our soul, of everything *spiritual*. (What we discover as already known is body again.)

No true cognition exists without the support of these three fundamentals, only we are usually unaware where we stand in concrete cases.

Spiritual discoveries reach their full significance only when properly related to the world of phenomena (bodily), permeated with our experience of our own self (in soul) and grasped with respect to their goal (in spirit). This means being so aware of these three fundamentals that it is always clear which reality we are dealing with.

* Rudolf Steiner, *Theosophie* (GA 9), Dornach: Rudolf Steiner Verlag, 1987; trans. *Theosophy*, London: Rudolf Steiner Press, 2005.

around us. What we perceive physically seems at first alien. It is a puzzle that arouses experiences in us and for which we seek meaningful solutions by means of our thinking. We also observe how we are separated from the world by our body and consciousness, but at the same time feel at home in it. Thus, when we are hungry, we feel the need to take substances into our body that are indeed alien to us but which we actually have a puzzling willingness to receive. These substances transform our body into its own substances, that is, into those which enable us to continue our bodily life and unfold a soul-spiritual life. Substances which alienate our body are excreted again. A similar process goes on in our soul with different kinds of pictures and percepts that we take in and which carry on acting consciously or unconsciously in our soul, changing us and constituting us.

Organs as gateways to the world

In order to feel at home in the world we need organs to interact with it in various ways. They open the way to acquiring a particular relationship to the world, but they limit our mind in other ways. For these we need different organs.

These organs include:

— sense organs, that is, physical, sensorial relationships such as seeing, hearing, smelling, touching etc.;
— life organs, which create the conditions of life in which *inter alia* alien substance is transformed into own substance and vice versa;
— soul organs, which create our soul relationships to other beings, be they human, plant, animal, atmosphere of a landscape, etc.;
— spirit organs with which we empathize with other beings and connect ourselves with them.

These organs, that serve the maintenance of life and cognition, are complemented by organs which are needed for expression and activity, such as limbs and organs of speech.

An organism comes from a germ

The physical basis of all the organs of our body comes from a common germ. Each organ is related to the others through the totality of this germ. They represent a special modification of the totality and thus together form the coherence that we call an organism, in which the original unity is retained. This is the opposite to a mechanism which is assembled externally from parts that originally had nothing to do with one another.

The physical body, as the basis of the organism, provides us with the awareness of being present, but in fact it only forms a limited aspect of the organs and creates their transparency to encounters in various ways. The physical make-up of, for instance, our eyes can then be fully regarded only as organs of sight if our environment becomes perceptible and experiencable in multicoloured pictures. Only the active connection of the human being to the world makes the organ an eye. Correspondingly, the structure of the ear prepares the way for the organ by which we can immerse ourselves in the sound world. Limbs are formed as organs in such a way that we can take part actively in the world and express ourselves. Inner organs, such as the heart, allow us to perceive our relationship between the inner and outer and always maintain a balance. In order to understand the essential nature of organs, it is important first of all to reflect on our relationship to the world that surrounds us.

Appearance and receptivity are separate in an organ

Each organ formation exists in a separation process of two complementary aspects of a whole: for example, sun and moon. So alongside the sun's aspect of the appearance of a sense percept is the moon's receptivity aspect that comes to meet it. Each makes the other into a whole. In consciously uniting a percept with the appropriate receptivity for it, what was at first puzzling becomes meaningful. Reality is experienced. An experience of absolute certainty happens in which a percept and its meaning become one. The particular orientation of an organ to the world, the special direction in which its activity is dedicated, emerges from the same spiritual source as what is perceived. They are two sides of the same reality.

Development and enhancement of human organs

Organs of the body are not just a feature of human beings; animals have them too. From the moment of birth, an animal's perception immediately encounters the receptivity that is caused by its own body, and which leads to its instinctive reactions. In human beings, self-consciousness enables us to some extent to distance ourselves from this. Our capacity to perceive with increasing discrimination and consciousness depends on our biography and on the development of our perceptual and cognitive faculties.

We are also capable of consciously and specifically developing new organs for perception in soul and spirit, but for this we need to perform the aforementioned separation process ourselves. This is important for practical work with biodynamic preparations. The value of such training lies precisely in strictly adhering to conscious separation and combination. If we are merely content with flashes of inspiration, it is not possible to tell the precise spiritual direction from which they came. If such inspiration arrives unexpectedly, then for true knowledge to arise, the separation has to be carried out afterwards.

Animals as representatives of particular attitudes to life

By comparison with human beings, the organization of each animal is specifically one-sided. As regards particular organs, each animal is better developed than the human being, because its instincts are impressed on its body from birth. It is precisely because the human body is less specialized from birth, and that we must to some extent develop our capacities in self-consciousness through learning and training, that we are able to use it more freely. We can recognize through being conscious of ourselves what is purely instinct for an animal. However, even if animals are unaware of their faculties, we can regard them as representing particular attitudes to life.

By accurately and pictorially studying the way animals behave, inwardly recreating each animal out of interest for its inner nature, we can empathize with its particular soul-world. We already do this each time we encounter an animal. This is particularly enlightening with vertebrates, but such empathy can be further developed by accurately observing everything expressed by the animal. However, we must observe them with real interest, pausing to appreciate the pictures that arise and avoiding merely reading our own soul experiences into the animal.

In the following, we shall sketch this process for two groups of animals, birds of prey and ruminant mammals, whose whole organizations are directed at polar opposite faculties on which their outward perfection is based. If we study their organization and behaviour by 'slipping under their skins' with painstaking observation, we will be able to see how they exist in the world, that is, see the precise world they inhabit. The pictures we receive can show us what faculties are required when in mind and body we adopt this or that attitude to the world.

Eagles are at home in the air. They have keen eyes. Thus they can see distinguish tiny moving objects (for instance, mice), swoop on them, grip them in their talons or beaks, tear them apart and devour them. Their digestion happens quickly and cursorily. An eagle's plumage is intricately differentiated, seemingly dried out.

Cows are associated with pastures and grazing, smelling with their moist noses and tasting, first with their saliva-moistened mouths, and then with their entire digestive system. They take their time over digesting and ruminating. Their eyes are not given to focusing on small points but rather are directed dreamily at generalities. The bodies of cows are rounded, soft.

The pictures arising from such contemplations are suitable for helping us to see which faculties are associated with the nerve-sense system (eagle) and which with the metabolic-limb system (cow), that is, how these systems are arranges cosmically. This can at the same time contribute to an understanding of the organ sheaths (see Chapter 6).

Organs of the organism of the earth

It is clear from the foregoing that *plant* life has a quite different orientation from that of animals and human beings. The physical appearance of plants can be experienced and understood by our imaginative perception, right into its seasonal life processes, through its order, beauty and radiance or adornment.

For aspects of the cosmos, see page 53.

The organ that enables a particular plant to thrive in the natural world is like a spiritual sheath round the plant, supporting it with the conditions and processes that together make up its habitat. Habitat seen in this way is an organ of the earth organism. It enables plants to receive in a specific way the light and mineral substances in their surroundings, to incorporate them into their life processes and to differentiate and consolidate them in altered form in the characteristic, material picture of their manifestation. Thus the seemingly physical orientation of the plant to the cosmos is, in the context of general plant life, like a gesture directed at the spatially infinite, spiritual periphery, like a picture of the special direction in the cosmos towards which the activity of the plant species is aligned.

We receive an immediate, vivid impression of how plants, growing out of the ground, open totally into the surroundings and express their way of doing this in their appearance. This impression includes a soul quality. However, unlike animals, plants cannot develop a soul nature in their own bodies.

The soil as a diaphragm of the earth

The earth's surface is a real organ, which — if you will — you may compare to the human diaphragm. (Though it is not quite exact, it will suffice us for purposes of illustration). We gain a right idea of these facts if we say to ourselves: above the human diaphragm there are certain organs — notably the head and the processes of breathing and circulation which work up into the head. Beneath it there are other organs. If from this point of view we now compare the earth's surface with the human diaphragm, then we must say: in the individuality with which we are here concerned, the head is *beneath* the surface of the earth, while

we, with all the animals, are living in the creature's belly! Whatever is *above* the earth, belongs in truth to the intestines of the 'agricultural individuality', if we may coin the phrase. We, in our farm, are going about in the belly of the farm, and the plants themselves grow upward in the belly of the farm. Indeed, we have to do with an individuality standing on its head. We only regard it rightly if we imagine it, compared to man, as standing on its head.

Rudolf Steiner, *Geisteswissenschaftliche Grundlagen zum Gedeihen der Landwirtschaft* (GA 327), Dornach: Rudolf Steiner Verlag, 7th edition, 1984; trans: *Agriculture Course: The Birth of the Biodynamic Method* (London: Rudolf Steiner Press, 2004), lecture 2.

Inner and outer in inverse relationship between plants and people

The situation is exactly the opposite with human and animal organs. Here a physical sheath is formed and the particular relationship with the world, for which the corresponding organ clears the way, arises at the soul-spiritual level.

What is a visible picture in the case of the plant, is a soul-spiritual process in the case of a human being, and what for a plant is situated at the spiritual periphery in the human being, takes physical form as a sheath, the form of an organ.

The animal as an assimilator of air and warmth

In the nerves-and-senses system ... the animal is itself. In its own essence, it is a creature that lives directly in the air and warmth. It has an absolutely direct relation to the air and warmth. ... But the animal cannot relate itself thus directly to the *earthy* and *watery* elements. It cannot assimilate water and earth thus directly. It must indeed receive the earth and water into its inward parts; it must therefore have the digestive tract, passing inward from outside.

With all that it has become through the warmth and air, it then assimilates the water and the earth inside it — by means of its metabolic system and a portion of its breathing system. ... The assimilation-process is of course, as I have often indicated, an assimilation more of forces than of substances.

Rudolf Steiner, *Geisteswissenschaftliche Grundlagen zum* Gedeihen der Landwirtschaft (GA 327), Dornach: Rudolf Steiner Verlag, 7th edition, 1984; trans: *Agriculture Course: The Birth of the Biodynamic Method* (London: Rudolf Steiner Press, 2004), lecture 7.

10. The Organization of Man and Animal in Relation to the Elements

Let us for the moment set aside all scientific knowledge on offer today about the four classical elements and images of them as aggregate states, and concentrate simply on a relationship in which we unfold our lives, namely to the qualities of earth, water, air and warmth.

With each sense percept, we experience, as if by touching, that there is an outside, and we think of it as material. In doing so, we easily overlook that this simultaneous experiencing and thinking creates our earthly self-consciousness.

Without our being able to hold on to a thought, we would not be able to speak of a solid element. Our self-experience is thus supported by the *solid element*, as is our body on the earth.

We experience ourselves in the quality of the *element of water* when we follow sequences of phenomena, immerse ourselves in them so to speak, and, while dreaming in them, or through the conscious activity of perceptive thinking, always adhere to the relationship that continues in the changes. In following random pictures we would not be able to keep afloat; we would sink. Becoming isolated, or losing ourselves in a constant stream of pictures, are dangers we are increasingly exposed to by the media.

Instead, we sense our self and our feelings — albeit gaining only an inkling of them —as independent from the fluid and solid elements. We absorb the substances of earth and water from without in order to maintain ourselves in a physical body in the process of change.

We may feel more related to the *element of air* than to that of earth or water. When we *enthuse* about an idea, we are engaged in a way of perceiving and experiencing that is spiritual. Such a way of perceiving may of course appear reflected

See Jochen Bockemühl (ed.): 'Elemente und Ätherarten als Betrachtungsweisen der Welt' in Erscheinungsformen des Ätherischen *(Stuttgart: Freies Geistesleben, 1977); trans: 'Elements and Ethers: Modes of Observing the World.' In* Towards a Phenomenology of the Etheric World *(Spring Valley: Anthroposophic Press, 1984)*

to us in dream pictures or more consciously in ideas, but it is not identical with them. The element in which we experience ourselves as being present is of the nature of air, which is certainly present somewhere, but is hardly perceptible. Precisely because of this it leaves room for experiencing the spirit that comes to life again in the phenomena; for example, in a landscape. The air seemingly borrows something of the fluid and concrete aspects, but in doing so immediately retires so that the mood of an atmosphere arises that connects us with, for example, a plant or a landscape.

However, we experience ourselves as identical with the *element of warmth*. If we become aware of it in what we are doing, then it is our own activity in grasping and penetrating the spirit that we feel united with, and to which we are committed.

Once we are aware of living with the elements, we can begin to understand what Rudolf Steiner meant when he pointed out that the human being lives rather close to the elements of warmth and air, but that solid and fluid elements have to be taken up and assimilated in forming the body wholly from within outwards and configured in, for example, bones, muscles, nerves and body fluids, in such a way that they can serve in life as tools.

To assimilate the fluid and solid elements, the human organism needs special organs that we shall discuss in the following.

Organs for assimilation of the fluid element (kidney-bladder system)

When we are thirsty we feel refreshed after drinking a glass of water. We feel the water permeate our whole body. With a second glass, we already sense that this is less the case. With each subsequent glass, we gain an increasingly stronger impression of reaching the limit of our capacity to take in more. In this process we experience how water is taken up immediately into the fluid circulation of our bodies. It permeates the whole organism without a special digestion process. The fact that it does not simply 'pass through' provides an initial indication of how important water is for human life, and of what is of especial importance in dealing with the water element, as well as its assimilation into our organism, namely enlivening, but also connecting and inwardly structuring.

In the kidney-bladder system, human beings have a bodily organ that is directed to sensing the subtle balance in the differentiations of the fluid organism within. By concentrating, and specifically excreting, substances the processes in the fluid streams are maintained and become independent so that a mobile inner life can come about. In that the fluid as it were 'volatilizes' — that is, materially disappears to leave room for superordinate processes — it forms the basis of the human capacity to develop a soul-life.

In order to unfold its inwardness, our soul requires that the body has no part in it, on the contrary it withdraws, creates a mobile clear space and bestows on it a certain stability. The urine excreted in this process is the dead counter-image (or phantom) of the individual astral body that transforms into a specific smell. As a direct result of the excretion process, this may become relatively freely mobile in soul.

Dogs have a good nose for such qualities. Many mammals mark the boundaries of the soul space they require in the landscape with urine. After resorption, this fluid accumulates in the membranous bladder prior to excretion. The organism is connected with the atmospheric-cosmic effects of its surroundings by the outwardly directed senses, above all of its head region. What this stimulates in the soul body is connected inwardly with the bladder. A young child, whose soul is still totally immersed in the atmosphere of its surroundings, therefore reacts to disturbances in those surroundings especially with its bladder. It first has to learn to control the fluid organism and, by doing that, maintain independence of soul.

Organs for assimilating the solid element

In contrast to the way we deal with the fluid element that flows directly into the human organism and has to be contained, differentiated, held in balance and purified, solid food must pass through many stages of breakdown and dissolution before it can be taken up into the organism. The digestive system is at the outset receptive to perceiving substances alien to the body. It is open to the wisdom that has entered into the ingested food substances and is thereafter put to work in the organism. It interacts perceptively with the substances, further changing them step by step. The still alien substances are first touched, mixed with saliva secreted by the organism during chewing in the mouth, and thus tasted more or less consciously. The tasting becomes duller during further processing and dissolution of these substances, helped by the secretions of the stomach, gall bladder and pancreas. To a certain extent the resulting pulp is now adapted to the organism. It is no longer aware of it as something separate, but *tastes* it inwardly.

This process brings the labile state of the substance to a peak in the *small intestine.* There secretion begins both inwardly and outwardly and the finishing touches are applied to the transformation of what is alien to the organism into what is its own. This acquisition of substance by the human organism proceeds in such a way that it no longer merely volatilizes the fluid element to enable its soul-life, but actually transforms it into a kind of state of inner warmth in which the person can unfold — that is, the substance is transformed into the element in which the individuality lives. This process contrasts with that of excretion via the large intestine and rectum. There, substances are concentrated, alienated, lifeless and thus to a large extent separated from the life processes of the earth. On this reverse path of substance towards consolidation, forces of the human I are released. Self-consciousness arises.

In cows, digestion and inner tasting of substance is more deeply seated, but what results is a dreamy picture consciousness and not self-consciousness. What is used for self-consciousness in human excretion is retained in the excreta of cows, and benefits the life context of the earth as 'I-potentiality.'   When Rudolf Steiner used this expression, he was clearly referring to a capacity for humus formation that could be taken up in the soil as the basis for a kind of individualization process.

After reading Rudolf Steiner's Agriculture Course we become aware of these matters and a contemplation of the qualities of excretions as manures can open totally new perspectives for insight into the nature of manuring and into what I, or individuality, means in the context of the natural world. For this we shall insert a comparison of brain and digestive system.

The *brain* is organized in such a way that in principle healthy human beings have the capacity to grasp concepts and thoughts spiritually and largely independently of bodily and soul influences. We have the potential for our thinking to resonate with the cosmos. The brain creates the individual space for this when we enter into the process consciously yet at the same time with self-restraint. We can look out on the world with various questions or with different ways of perceiving or orienting ourselves. People who are disposed to it may think materially, but they are not forced to only think in a material way about the world! The world has still other aspects, some of which are considered in this book.

Animals have not yet the capability of accessing the spiritual in freedom. They are directly connected with spirituality, from which they have developed, through the organization of their brain. Their instincts have arisen directly from the forming of its body. As a result, each animal realizes a particular constitution of soul in the context of the natural world. It lives in a fixed picture of interrelated images of its environment, not in a world of separate, freely-manipulable thoughts and objects like that dwelt in by human beings, who can make new connections between such things in thinking and thus be creative.

In other words, the brain inwardly produces the connection with the world of the spirit (cosmos). In human beings it generally opens a sense for spiritual matters; in the animal, a vision in a particular direction, into a 'world-view' complete in itself, which it immediately acts on and from which it cannot escape.

We can regard the intestines, with their excretory function within the metabolic system, as the opposite pole to the brain. Just as at one pole, in the brain, pictures of sense impressions are 'digested', giving rise to soul faculties, so at the other pole, the intestines, food substances are digested so as to form the species-specific body on the basis of the foreign substances, and develop physical forces. At the level of the solid element, what human beings excrete is largely exhausted, but, by contrast, in herbivores this is not the case. Cows, living only in their soul-world, produce a manure with their particularly thorough digestion that contains a substance that reflects the counter-image of their soul-life without this resulting in self-consciousness. Cow manure is thus an ideal excretory substance for the individualization process that is required in agriculture to enable the healthy development of food plants in the context of the organism of the earth.

Organ systems for making connections: the nerve-sense system in relation to the element of air, and the metabolic-limb system in relation to the element of warmth

In human beings, as in animals, two major organ systems are polar opposites of one another and thus create the conditions for life processes that support soul-activity to unfold in an earthly body. These organ systems, which hold together and regulate the other organs, are superordinate to them and enable a bodily cosmos to arise. They are organs that in manifoldness contain the mobile unity of the soul.

One is the nerve-sense system with the *brain*, in which perceptions of all kinds, coming from all directions, collect together. The brain is receptive to letting such percepts congregate in the mind. This happens dreamily in animals. Human beings can consciously turn percepts into thoughts one by one.

It is the bony skull that creates the free space for the activity of the brain as the central organ of the nerve-sense system. The brain floats in the intracranial fluid and is thus almost completely freed from the influence of gravity. Brain tissues are living, as are those of the rest of the nerve-sense system, but they are constantly on the verge of death. This applies all the more to the tissue within the skull, which essentially comprises mineral-like deposits, although it remains connected with the life processes.

In other words, ensouled and living substance is brought into a mineral state in this organ system in such a way that cosmic forces are placed at the organism's disposal. Various degrees of dreamy consciousness arise as well as, in the human being, awake self-consciousness.

The second of these organ systems is centred in the abdomen. It includes the activity of all organs that serve the life processes and the synthesis and degradation of substances. This organ system regulates their interaction so that bodily activity can happen.

The organ that enables the central function, as the skull does for the activity of the brain, is the *peritoneum* which, from within, clothes and delineates the abdominal cavity and part of the metabolic organs. It forms the free space for a substance that cannot be fixed, that is in constant renewal and forms no structures. Considered outwardly it is a flexible fluid, a kind of lymph, that mediates between the various functions of the organs of the abdomen, keeping them together in a living cosmos. This is particularly puzzling for science. Surgeons generally think that the peritoneum, together with the fluid of the peritoneal cavity, promotes the regeneration of organs in all parts of the abdomen.

The cells of the actual metabolic organs, in contrast to those of the nerve-sense system, are in a constant state of renewal at different rates. An intensification of the functional tendency of the abdomen can be seen in the creation in the womb of an entirely new organism which becomes independent.

In the metabolic-limb system the spirituality of the cosmos is largely united with substance to the extent that it individualizes into a personality, or becomes one-sided in the being of an animal. The capacity of the bodily organism to grow, regenerate and maintain itself in balance is also localized here, but as with the nerve-sense system, the metabolic-limb system permeates the entire human organism.

Between these two key systems lies a third that enables the activities of the first two to flow into one another so that life as a unity, and soul-experience, is possible. This is the rhythmic system. It includes as single organs the heart, lungs and blood circulation. The blood connects all organ systems. The diaphragm delineates the rhythmic system from the metabolic and participates in conveying the life processes between the organ systems.

In view of the fact that human beings experience the world as something separate and can at the same time feel at home in it, the rhythmic system has a special position in them. Self-consciousness does not happen in animal life. In animals, the relationships between the nerve-sense and metabolic-limb systems are already fixed from birth onwards according to the particular animal. Its soul-tendency (see examples of cow and eagle on page 118f, that is, its astral or soul body, is formed in a particular way, more or less firmly imprinted on its physical body. The mediation function of the rhythmic system achieves no independence, because its consciousness is directly linked to the formation of its organs. This is most obvious in the structure of the limbs. As Goethe put it, 'animals are taught by their organs'.

This is only partly the case with the human being. What sets us apart is precisely the capacity we have to intervene between perceptions and will impulses in such a way that we ourselves have to learn how to master the connections between them. This is why a small child is at first so helpless, but later it gradually learns to instruct its organs itself.

Thus, through self-observation, we increasingly learn to understand not only what is happening in our organs, but also a comment that Rudolf Steiner made in his Agriculture Course that can be briefly summarized as: in the head pole we are dealing with the earthly substances and cosmic forces, and in the metabolic-limb system with cosmic substances and earthly forces.

Healthy people are aware of their bodily organism only in so far as it conveys perceptions and enables activities. This means that in the body we can develop our soul-life according to our innermost nature with thinking, feeling and willing, and it is capable of serving as an instrument of our perceptions and intentions.

Summary

In the preceding study of human and animal organization in their relation to the elements we have emphasized two pairs of organ systems in particular.

1. The kidney-bladder system and the gastrointestinal system

The *kidney-bladder system* serves the containment, refining and volatilizing of the fluid body with its growth and degradation processes. It forms the basis for the creation of a free space inside the body for the unfolding of the soul in which animals live dreamily in a way that accords with their natures. The *bladder* provides the whole system with the physical support for the containment, but it is also sensitive to cosmic forces acting from the periphery.

The gastrointestinal system serves the uptake and excretion of earth-related substances; at the culmination of this process, in the small intestine, begins the transformation of substances alien to the body into its own. Here, the small intestine is the key organ in which the liquefied earthly substance from outside most closely resembles the organism. There it can be absorbed through the intestinal wall and be transformed, through a kind of inversion, to make soul and spiritual life possible in an earthly body. This means that the substances change to *aeriform state*, permeable to the soul, and finally to a warmth state in which the being can feel united with its body.

2. The head or nerve-sense system and the metabolic-limb system

The *head* or *nerve-sense system*, serves to gather all sensorial perceptions together. In it substances are brought from the living organism to the verge of death so that the spirit living in the being can become more or less conscious. This is the basis of the human capacity to think separate things consciously, yet also participate with imaginative perceptive thinking in the totality of the spiritual world. With substances becoming earthly and mineralized in this prominent organ, *cosmic forces* are much freer, for forming the body initially and, later, for consciousness. The physical support for this is the skull which, of the four organs, coincides most with the outer physical form of the body.

The *metabolic-limb system* serves to bring together the metabolic functions of all organs to enable the whole organism to be physically active. In the free space created inside the body by the *peritoneum*, with its highly vital fluids, *substance* is closest to a living-cosmic state. In it is an active receptivity to the effects of the cosmos exists that accords with the particular being out of which that being is active on the earth.

11. The Effect of Plant Substances on the Human Organism

The healing relationship of the preparation plants to the human organs provides a key to understanding their role in agriculture. Substances built up and formed by plants, into which particular cosmic effects have entered in the way described, are not only an expression of a picture created — that is, of the atmosphere that radiates out and touches and nourishes our soul (cosmic nutrition) via our senses — but are also used as medicines.

The outer form of the plant substances, which in growth and differentiation have reached an end, can be removed. They are converted into a fluid state in various ways, either by a pharmaceutical process or in the human organism, so that they can act in the way that is imprinted on them. The organism first experiences this as alien effect that it must overcome in order to summon up the healing effects in itself.

If the medicine is carefully chosen, prepared and administered, then the process of overcoming this specific foreign substance in the human organism awakens forces which are too weakly developed in a particular organ. Thus the substance, through its relationship with the organ in question — that is, through its own nature finds the way to the place where its effect is needed. The organ region of the plant from which the substance is taken directs it to the corresponding organ region in the human being. Generally we can say:

— root substances affect the nerve-sense system;
— leaf substances affect the rhythmic system;
— flower and fruit-seed substances work on the metabolic-limb system.

Healing effects of preparation plants

While studying the ways the preparation plants are formed, we discovered characteristic gestures that we tried in each case to relate to the elements. We then moved on to considering the human organs that are related to the elements in a way that is the reverse of plants. Likewise, contemplating the way these organs work led to an inner perception that we can experience as a gesture, as a particular orientation to the cosmos. Mindful of the corresponding reversal, it is now a question of discovering the relationship between the form of the plant and the soul-spiritual faculties of the human being. In order to individualize the metabolism of the earth for the cultivation of food plants, typically it is mostly the flower substances from the plants studied that are used, or, in the case of oak, the substance that is in the most complete form.

Yarrow creates a free space for the development of the soul-element
Through its fine network of roots, yarrow senses the minerals and sends them upwards with the flow of absorbed water. This guides the greening organs into a refining process in which traces of the fluid element still remain in the pinnation as far as the delimitation of the whole leaf. The same limiting gesture of holding back reveals itself in the inflorescence, in the way the fruit is held together during ripening, in the stem becoming hardened and in the tartly aromatic taste that increases towards the flower heads but is restrained. Although an atmospheric element flows outwards in flowering, it does not extend to the release of warming scent.

The function of the kidney-bladder system shows the same inward gesture in that it maintains the fluid body — that is, produces inside it a condition of airy refinement that creates a free space for the soul to unfold.

Chamomile stimulates the dissolution of alien substances and their transformation into the body's own substances

Chamomile takes its supply of mineral nutrients from the ground with a strong-smelling, thin, branched tap root and then grows away from the earth, constantly transforming and refining itself. It everts itself in flowering and finally exudes itself in a warming scent and by scattering minute seeds.

On the boundary of the small intestine, in a correspondingly in-turned gesture, a process of dissolution, uptake and refining of food substances takes place. This food from the earth is transformed into ensouled warmth-substance permeated by the human being — that is, into a substance through which the human being can live in a physical body.

Oak has a constricting effect on the blood and helps to bring about consciousness in the nerve-sense system

In contrast to chamomile, which constantly grows away from the earth, the life processes of oak constantly raise the earth up somewhat at various levels. While the substances of the bark approach the mineral state, a simultaneous counter movement ensures that the way is always kept free for new shoots.

Consciousness arises on the way to dying in the correspondingly in-turned functional gesture of the nerve-sense system.

Dandelion brings about an enhancement of life processes and restores the organism

With its gesture of swelling growth close to the ground and, at the same time, its flowers opening to the spring sun, dandelion is an outer picture of the core of the metabolic system, in that the functions of metabolism, turned inward, form a living cosmos which serves to build up and renew the human organism.

Stinging nettle helps to maintain the balance between metabolic and nerve-sense systems

The gesture of nettle speaks of conveying from above downwards the ordering forces of the cosmos, and from below upwards the rising, ordered life processes.

Correspondingly, the heart-circulation system mediates the balance in the human soul between inner and outer and the diaphragm is the physical representative of the boundary at which the metabolic processes interrelate rhythmically with nerve-sense system functions.

Valerian promotes ripening and harmony in the metabolism

The root sap is used in human medicine. Applied in the proper way it works on sleep consciousness to harmonize the soul-body.

Horsetail promotes transparency for the spirit

Horsetail's intricate form expresses how the plant, supported by silica, creates a picture of an aeriform, atmospheric element by outwardly depositing its cosmic nature.

This process, that becomes a picture in horsetail's appearance, is turned in the human organism into a capacity for forming the bodily aspect needed to promote permeability to the soul-element in the various organ regions. This happens for perception in the senses and for the expression of our inner life that flows into outer activity in the bones and limbs, but it is also important that, through the quality of silica, faculties of inner perception arise.

12. Organs of the Landscape and of Soil Chemistry

Aspects of understanding agriculture as a kind of individuality with its own organs can be derived from the relationship of healing plants to the human being.

The preparation plants serve to support the individualization process aimed for in biodynamic farming. In the metabolic process of the earth organism too, we can speak of two different directions of organ activity that should be developed and supported; namely organs that work out of the landscape surroundings, and organs that create an appropriate receptivity in the chemistry of the soil. This results in the interaction of the earth's life processes becoming strengthened in such a way that, from out of the entire nature of a farm, they help cultivated plants to thrive — which of course can only thrive in natural surroundings that have been steered and shaped by people.

Organs of the landscape

If we want to find again, via an inversion process, what comprise the human organs as orientations to the world, in the outer pictures of the surroundings in which we live, it means discovering corresponding tendencies in the real appearances, pictures and processes of the healing plants with their habitats in the landscape.

In this way, the forms of the healing plants become pictures of soul-spiritual processes in the human being. With these in mind we can speak of organs of the landscape that enable these plants to thrive. They are connected with the atmospheres of the habitats discussed and in each case expose the growing plants to effects from the cosmic periphery.

Revitalizing the earth

The water in the earth can be revitalized by mineralization processes, but this does not give rise to any revitalization of the earth because things merge into one another in the element of water. No interaction can arise that is necessary for the life of the earth. A characteristic feature of a life process is that, as with human nutrition, a constant interchange exists between inner and outer, between growth and degradation. In the merely fluid element there is no inner and outer. Everything interpenetrates everything else. It is precisely for this reason that the support of the earth element is needed.

Direct revitalization of the earth can be achieved by starting with organic fertilizers and turning these into a condition that has an organizing and revitalizing effect on the earth. In order to obtain the appropriate surplus for the formation of food, organs have to be laid down between which the interactions that correspond with an organism can take place, so as to develop in them a kind of reason in the life of the earth.

Organs of soil chemistry

All plants obtain their specific receptivity for these effects through the mineral substances with which they are connected in certain ways. To understand this, we need to characterize a chemistry of life processes in the way that is meant here. Unfortunately this can be dealt with only briefly within the confines of this book.

The chemical properties of a substance are activities that are opposite to the properties of solid matter. The latter have to withdraw when the former are active. As a result, certain new properties reappear. Thus chemical properties are always processes with particular directions, whereas the properties of matter represent processes that have come to an end. Rudolf Steiner referred to the chemical elements — above all when discussing life processes, for example in his Agriculture Course — as 'bearers' of different effects. We understand these as working principles that express themselves in polar opposite ways in the natural world. If we seek the common factor in the multiplicity of effects of an inorganic element interacting with others, we will gradually come to an inner perception of the particular principle that reveals itself in a certain reciprocal relationship between appearance and effect.

Each chemical element is active in the earth's organism in a particular cosmic direction. If it is taken up by the life of the plant, it conveys to the plant a capacity to develop its life processes in that particular direction.

The elements carbon, nitrogen, oxygen, hydrogen and sulphur form the basis of living protein. In the context of nature they appear individually or in compounds with other elements in more or less stable forms to which they always tend to return. Plants have no access to these stable forms. Only when oxygen is combined with the other elements of protein can it be taken up into the plant's processes and become organs for receiving cosmic effects.

The elements of proteins become plant formers

— *Carbon* occurs naturally in the earth in stable form as coal, graphite and diamond. Revitalized by oxygen, it 'volatilizes' and enables plants to take on their physical structure via a rudimentary 'carbonization'.
— *Nitrogen* is stable in the gaseous state. Revitalized by oxygen it tends to go into solution and can then enable plants to acquire an aeriform nature — that is, it becomes permeable to spirit form that, as a picture, is filled in by carbon.
— *Sulphur* manifests itself in various forms. To be accessible to the plant it needs to be in the form of sulphate, that is, in solution combined with oxygen. In compounds with the other elements, it conveys the spirit form connected with the seed into the life processes of the growing plant.
— Likewise, *hydrogen* is taken up by plant life processes not in its elemental form as a light gas, but via hydrogen compounds, and it bestows on them the property of volatilization.

These five protein elements form the actual body of a plant. They work together in different ways in each species, like organs of an organism. In each case their function is to give direction to the processes of formation.

This organism of life acquires its connection to the earth by means of other elements, namely *potassium* and *calcium* on one hand and *silicon* on the other, together with *iron* which has an intermediate position. None of these elements occur in nature in elemental form.

— *Potassium* is available to plants in dissolved salt form and conveys the capacity to give a certain solidity to the already burgeoning fluid growth and to maintain that solidity, in keeping upright and gradually giving itself structure with carbon.
— *Calcium* guides these processes further into a general ordering solidification.
— *Silicon* combined with oxygen enters plants in the ascending flow of liquid, like potassium and calcium, but, as silica, it is very insoluble and hardly participates in plant chemistry. Just as air makes room for experiencing atmosphere, silica enables the life processes of the plant on the way to earthly hardening, to become transparent to the specific spiritual aspect that is at work in it.
— *Iron* is taken up by the life processes of plants and enables them to keep their relationship between inner and outer in balance. It energizes, equalizes, mediates.
— *Sulphur* supports *phosphorus*. Its role is in promoting healthy fruit formation in plants and soil.

Interactions of the principles of action in the preparation plants

The aforementioned properties of the chemical elements are individualized through integration into the life processes of the preparation plants. What enters in this way into the particular substance formation of these plants can be returned directly to the soil life via compost or horsetail. The remainder of this section summarizes how Rudolf Steiner characterized this process for each preparation through the interaction of the aforementioned action principles of the chemical elements.

Yarrow, by comparison with other flowers, has almost miraculous powers. It develops its sulphur activity primarily in the potassium process. Sulphur is combined in it with potassium in a way that enables it to radiate its influence over large quantities of material.

Chamomile is capable of binding together necessary substances for plant growth. Sulphur, potassium and calcium interact in it to give compost the potential to absorb so much life that it is able to share it with the ground in which other plants are growing. This can contribute to eliminating damaging fructification processes from plants.

Stinging nettle is the greatest benefactor of plant growth. It extends the flows of potassium and calcium with sulphur which everywhere orders and assimilates the spiritual element. Apart from this, it has a kind of iron radiation corresponding to the iron in our blood. It sensitizes the compost and can bring 'reason' to the soil. This enables it to individualize itself into the plant that we want to grow in such a way that it develops a consistent ability to nourish.

Oak needs calcium in the structure of its bark for a controlled reduction of the etheric tendency to rampant growth. Through compost it enables the soil to work prophylactically in the prevention of harmful diseases in plants.

Dandelion can attract silica out of the entire cosmic surroundings and bring it into interaction with potassium, as well as passing on this capability to cultivated plants via the compost and the soil, which results in them becoming sensitive to influences from the immediate and wider environment. Such influences could otherwise pass by an insensitive plant. It would not be able to make this available for growth.

In relation to what is called phosphorus-substance, *valerian* enables compost to behave in a way that intensifies the ripening process if the flower juice, diluted with warm water, is added.

Horsetail has the special property of being able to deal with silica such that its entire form is expressed in delicate structures in mineral, opaline silica.

13. The Production of the Biodynamic Preparations

If we empathize with the living connections in the natural world in the way described, and learn to recognize what potential there is in the landscape for forming the organs of the earth and the way in which they are related to the chemistry of the life processes, then we can put these things into practice.

Creating organs of the landscape

Very definite organ formations are necessary for developing a farm as a healthy organism that will enable the plants cultivated there to flourish. Using the human organization as a model, we have tried to show, via the healing effect and the way substance is formed, to what extent the preparation plants can play a key part in individualizing a farm. If there are places where, for example, yarrow thrives, it is an expression of the fact that what corresponds in the human being to the kidney-bladder system is active atmospherically as an organ in the farming there. The same applies to each of the plants discussed, or each of the organs. If, however, one of the plants identified is missing, then the place is lacking the corresponding organ. It is not sufficient then that the plants are merely grown in beds. In addition, their surroundings need proper attention too. This has to be taken into consideration in the layout of the farm.

Where the plants are missing, if proper habitats are created in which they develop independently and particularly well, then organs are created whose specific effects are combined into an individual whole. If we pay attention to this in the layout of the farm then the concrete landscape situation will unfold together with what is living in the people who, in carrying out their work, develop a spiritual awareness for the individual preparation plants or the organs that correspond to them.

Forming organs in the living organism of the soil

A second way of making the activity of the preparation plants effective in agriculture takes the reverse direction, but is closely related to the first. As in human medicine it reckons with what comes from the substances that are formed by the plants with the help of the effects of the surroundings and the chemical elements.

In other words, the organs in human and animal bodies correspond to those in the farm landscape that send their influence from the spiritual periphery, although always in close connection with the earthly aspect. In this way they base themselves on particular locations just as the spiritual formative potential of a plant species is based on its physical seed. The effects of these landscape organs go into the *substance formation* of the plants.

The developmental pathway of a plant substance in the seed comes up against an inner boundary. It lets it free itself from environmental effects and gain the capacity to co-determine the future formative direction of the plant in a new combination with the earth.

The part of the plant that outwardly reaches completion, and in fading *falls* out of the individual plant, in fact enters into the general vitality of the soil through rotting and there forms the basis for the earthly growth of plants as humus.

The plant substances that have come to an end can be rendered active in the soil through compost in such a way that they do not disappear into the general vitality of the soil. To stimulate a differentiation of the organs of life in the vitality of humus, it needs to be digested in a particular way. In order for the particular cosmic tendencies of the individual preparation plants to transform themselves so that the appropriate receptivities are enhanced, we need something to contain their specific action and thus enable each transformation process to take place in the right way. For this we use organ sheaths that originate from animals.

Organ sheaths promote the specific transformation of plant substances

Insight into both the functions of the human organs and the healing properties of the preparation plants indicates where we need to be looking amongst the animals, that is, where the corresponding human organ functions are developed for the desired one-sidedness, so that what delimits them from other organs can be used as sheaths for the preparations.

In this context the importance of the *cow* as a ruminant is enlightening, that is, it has a special digestion, and as a domestic animal comes closer to people in a certain way. For these reasons, most of the sheaths of the cow are used. Another animal is used only for its bladder, namely the stag.

The stag, with its bony antlers formed entirely outwardly from within, is particularly sensitive among animals for what is going on in the atmosphere. His sensitivity is especially high during the annual regrowth of the antlers. The bladder, which in the life of the mammalian organism is a reflection or phantom of the astral body, is here used as a sheath for containing the *substance of yarrow* and converting it into a specifically active condition that is the opposite of seed formation.

The bladder, as a *sheath,* is intended to promote the transformation of yarrow substance into germinally active substance so that it induces spiritual structuring in the process of nature via the fluid element.

The cow's *small intestine* is the organ in which the labile state of substances reaches a culmination, and where the transformation of what is alien to the organism into what is its own begins through inward and outward secretion. The development of *chamomile* starts from the deep penetration of the mineral soil. What then appears in the rising flow of chamomile formation can be regarded as an outer picture of what happens inwardly in the metabolic process of the small intestine, namely the transformation of the solid and liquid elements into those of air and water. From this point of view the small intestine is the most appropriate sheath to contain chamomile during the rotting process if it is to change into the germinal state necessary for soil formation.

The bony form of the *skull* envelops the brain which floats in the intracranial fluid and is thus lifted out of the gravitational influence of the earth. Here, at the verge of death, formative forces for consciousness can become free. Turned into the outer picture of plant substance, this gesture corresponds to what appears in oak bark formation where, during the deposition of mineral calcium substance, forces are set free for new shoots. This recommends the skull for the sheath of the oak bark preparation in which this process develops most typically.

The *peritoneum* is especially well developed in cows. It comprises the serous membrane which lines the wall of the abdominal and pelvic cavity as well as part of the abdominal and pelvic organs. In between the organs and the wall of the peritoneum, the double-layered mesentery and the omentum are situated. Like an inversion of the skull, the peritoneum envelops a frequently overlooked hollow space which bounds all organs connected with metabolism and has a certain sensitivity to what is happening in and between them. There is an inner correspondence with the form of the *dandelion*, an especially vital plant in which in spring both flowering and vegetative growth appear from the surface of the ground while the root continues to branch in the soil depths. Its flowering lies between the root's sensitivity to the soil minerals and the delicate structure of the pappus.

Production of the *stinging nettle preparation* requires no animal organ. The ordering, stimulating force that during nettle's development has gone as a gesture into the picture of the phenomenon of substance formation is directly sunk into the humous soil, only surrounded by peat, a poorly rotting, isolating plant material.

Rudolf Steiner compares the *layer of the earth*, out of which plants grow, with the diaphragm that mediates between the metabolic region and the upper region of the organism. He is referring here to an outer process which is connected in the human being with the rhythm of breathing and in which atmospheric life is introduced into soul-life.

In the case of *valerian*, we are dealing with the juice of the flowers and retaining the scent through a fermentation process.

An additional procedure comprises exposing the ensheathed preparation substances to the influences of the seasons in specific ways, to the weathering processes of summer and the soil forming processes of winter.

The preparations are harvested at the peak of plant development

The substances used for the preparations of yarrow, chamomile, dandelion and valerian are harvested at the flowering stage. The flower substances are then at their most finely differentiated. With them the inner nature of the plant, its being, has reached its summer peak. On the way to manifestation, its substances are most outwardly directed. As a result, the life processes of the individual plants withdraw. Next follows fading and disintegration, which is guided in a particular direction by preparing them in the sheaths.

In the case of oak and horsetail the substances are harvested at a stage in vegetative development where the mineralized form has reached an end point. The deposition process in the outermost layer is similar to the formation of flower substances in that in both of them we are dealing with substances that have come to an end, albeit at a different level. In the case of stinging nettle we gather the fresh greenery when it is more or less tending to flower.

What did Rudolf Steiner actually mean when he emphasized in his Agriculture Course that with these measures he always maintained a living connection with the natural world? Combinations of the kind he describes are not found in nature. They are a completely new creation. However, if we approach an understanding of the aims of healthy agriculture in the way that we are trying to do here, we can recognize the way in which these preparation methods support the present life of the earth as it continues onwards from the past, but now stimulating in it something completely new, germinal, that through a kind of individualization, according to the model of the human organization, leads, with the appropriate reversal, to the future life of the earth.

A way of working that takes into account the whole context of life and into which people connect themselves responsibly with the earth's life processes, contrasts with working with isolated substances that are separated deliberately from the context that brought them into existence so that they can be worked with in a mechanical manner. Thus the effects of the preparations cannot be understood in the sense of modern, conventional analytical chemistry, precisely because they are linked to the contexts of nature and continue to reflect such contexts. It is not only each human activity that has an effect on, and changes the life processes of, the landscape but also the same applies to every thought harboured in our minds. This is because such thoughts determine the way in which we work. The more we succeed in raising to consciousness the conditions of a place in the way encouraged in this book, using them to guide our necessarily one-sided, purposeful activities, the more the requisite balance will be provided by the use of the compost preparations in connection with the overall layout of the farm. The effect of the preparations becomes an inner experience, in that we carefully interrelate in picture form all that is observed in the way described and constantly remake this picture from the phenomena that we are dealing with in each case. If we build up the picture for the love of it and allow this to work repeatedly on conscious experience, our own active attitude of mind will develop in parallel with the influences promoting fertility in agriculture. It is not easy to summon up the effort of will necessary for this, but it will bring new fulfilment to our lives.

14. Practical Details of Producing the Preparations

Farmers, horticulturists and scientists have long worked with and applied Rudolf Steiner's indications for the preparations. As a result, traditions have arisen regarding their manufacture. In this respect we would like to draw attention to a published introduction to the subject. It is also worth contacting a local biodynamic preparations working group.

We hope that our book will contribute to a deeper understanding of the Agriculture Course and enable people thoughtfully and factually to examine these traditions for themselves, and develop the right attitude of mind. This will allow the preparations to work in a way that is totally different from that of trying to achieve, particular superficial effects unconsciously out of the predominant contemporary view, for example, a quick rotting process or a greater yield in the short term. The opposite of this requires that the aim be to develop 'reason' in the soil — a totally different approach. It means experiencing the conditions of life in which we are intervening in order to bring quality out of the whole.

A summary follows of how the preparations are produced today, based on the indications given by Rudolf Steiner in his Agriculture Course.

Christian von Wistinghausen, Wolfgang Scheibe, Eckard von Wistinghausen, Anleitung zur Herstellung der biologisch-dynamischen Feldspritz- und Düngerpräparate, *Stuttgart, 1996; trans.* The Biodynamic Spray and Compost Preparations – Production Methods, *Stroud: Biodynamic Agricultural Association, 2000.*

Yarrow preparation

The bitter-tasting inflorescences are picked, dried like tea and stored. In the following early summer the plant material is moistened with the pressed juice or tea from fresh yarrow and packed into a stag's bladder.

The stag's bladder filled with yarrow flowers is hung in a sunny place and exposed to atmospheric influences. In the autumn it is taken down and buried in the ground where it is left to overwinter, that is, at the time of inhalation of the earth. In spring, the finished preparation is dug up and carefully stored until it is added to compost.

Chamomile preparation

The white and yellow flower heads of chamomile are gathered, left to dry or wilt and immediately surrounded by the animal sheath. As a sheath we use the small intestine of a cow.

Sections approximately twenty-five centimetre long are cut from the small intestine and packed with chamomile flowers. Like yarrow bladders, these little 'sausages' are either hung in a sunny place exposed to atmospheric influences and buried in the autumn, or they are buried in the autumn without hanging out in summer. In either case they are left in the ground to overwinter.

Stinging nettle preparation

Nettles that have reached the beginning of flowering are cut early in the morning and left until the afternoon to wilt. They are then packed into a hole in the ground (without an animal sheath) in such a way that they are completely surrounded by a layer of peat. This is then covered with what was dug out of the hole. They are left buried in the living soil exposed to the effects of the whole course of the year.

Oak bark preparation

The outermost bark layer is removed, ground up and packed into the brain cavity of the skull of a domestic animal.

The foramen magnum, the hole in the back of the head, is finally closed with pieces of bone. Unlike in the cases of yarrow and chamomile, the skull is not buried in richly humous arable soil but instead in marshy conditions where fresh rainwater and snow meltwater flows periodically through a layer of mud made of decomposing plant material. This could be a vessel placed under a rainwater outflow. In the spring, the oak bark, now earthy, is taken out and is ready for use as a preparation.

Dandelion preparation

In early spring, dandelion flowers that are still firm in the middle are gathered from pastures and meadows and dried carefully so that they retain their beautiful yellow colour.

Bovine mesentery is used to ensheath them. In practice, either the actual mesentery or the greater omentum is used. The mesentery is a piece of connective tissue from which the intestine hangs, and by which it is kept in position. In ruminants the greater omentum envelops the rumen like an apron and extends to the loops of the small intestine. Although these organs look different, they are part of the more extensive peritoneum and are related in function to each other.

The dried flowers are moistened with the tea or pressed juice of dandelion greens and packed into a piece of mesentery or omentum to form a spherical shape, and then tied with string. The bag is either hung outdoors in the summer like yarrow bladders and buried in the autumn, or buried in the autumn without hanging.

Valerian preparation

The inflorescences of valerian are picked, finely shredded and pressed. The emerging flower juice is stored in bottles with rubber caps. It becomes brownish and smells strongly aromatic. The liquid valerian preparation is added to hand-hot water and sprayed over the completed compost heap.

Common horsetail preparation

This medicinal plant is not one of the compost preparations but is of particular importance in biodynamic cultivation as a prophylactic treatment against fungal attack.

The green above-ground shoots are gathered and dried in late summer when the greenery is firm and fully mature and the silica content is at its greatest. A tea is obtained by boiling the plant material for a long time and, after dilution, is sprayed regularly on the ground.

Appendix

Preparation plants in different climatic regions of the world

People often ask whether there are substitute plants which may be used when the indicated preparation plants do not thrive in a region with a different climate. This can be answered by saying that there are almost no countries where most of the preparation plants do not thrive. Where one or another does not thrive, it is largely attributable to people not having sufficiently investigated how to adapt the plants to the climate where they are being raised. For example, close to the tropics there can be problems with the flowering rhythm. In Johannesburg we were once shown valerian that would not flower, but on closer examination a few examples were found flowering in the shade provided by shrubs. We should question the view that only indigenous plants should be used.

Rotting processes happen in the same way in different parts of the world if we create the conditions for a good transition of decomposing substances into soil for cultivation. Cultivable agricultural soil means fertility for food plants that would not thrive but for our help. We need only think that agricultural arable soil means infertility for many wild plants. Most cultivated plants originally came from other countries. In the few cases where they were indigenous from the beginning they became, like all other cultivated plants, alienated from nature that was left to its own devices.

It is, therefore, better first of all to stay with the preparation plants indicated and if possible create places where they thrive, even if to outer appearances they deviate from the way in which we experience them in Central Europe. The resulting efforts lead to discoveries that help us better understand our own locality. Where, despite everything, it is necessary to look for substitute plants, it is important first to contemplate the model plants and their specific roles in organ formation in the individualization process of a farm.

From the point of view of what we have presented here, it is understandable, however, when people find it unsatisfactory working with imported preparation plant material. But for those starting off, this is always a suitable option.

Central European oak species and their suitability for the biodynamic compost preparation

There are many oak species in the world, especially in the Northern hemisphere. In Central Europe only the following are indigenous:

Pedunculate oak (*Quercus robur* L.)
Sessile oak (*Quercus petraea* Liebl.)
Hybrid oak (*Quercus x rosacea* Bechst.; *Quercus robur x Quercus petraea*)
White oak (*Quercus pubescens* Willd.)
Quercus x calvescens Vukot (a hybrid of white and sessile oaks)

These species or hybrids are very similar to one another. Linnaeus did not distinguish between the first three and thus called them all *Quercus robur*. The distinctions were only introduced by later taxonomists. No distinctions are made as to origins amongst these species when it comes to making medicinal preparations of oak bark. Likewise, 1, 2 and 3 are used in the production of tanbark.

White oak is smaller, but otherwise very similar. Its distribution is primarily in Southern Europe, but also in southern Central Europe in vineyards (very frequently in the hybrid form). It is also suitable for use in the preparation.

Turkey oak (*Quercus cerris* L.) and holm oak (*Quercus ilex* L.), South-East European species which also grow in Switzerland and Austria, are very different in character. The red oak (*Quercus rubra* L.), introduced into Central Europe from North America long before Rudolf Steiner's time, is also very different in overall growth.

In principle, both pedunculate and sessile oaks can be grown in non-European countries. In warmer climates, especially in the Mediterranean, the bark of white oak is certainly worth considering for the preparation.

The taxonomy of officinal valerian, the group Valeriana officinalis s.l.

According to recent research the *Valeriana officinalis* group in Central and Southern Europe is divided into four morphologically defined basic types, which are not, however, taxa in the narrower sense; transitional forms mediate between these types:

1. Basic type *exaltata* includes tall plants with outstretched inflorescence when the fruit becomes ripe. Pinnate leaves with long petioles, a modest number of fairly wide leaflets (end leaflets narrower than the side leaflets) that are strongly indented. No hair on stems and fruit. Small flowers, medium sized fruit. Plants form rhizomes, no runners, thin roots.
2. Basic type *collina* has small plants with small inflorescence. Pinnate leaves, short petioles and a large number of narrow leaflets, smooth-edged to shallowly serrated, few indented. May form subterranean runners.
3. Basic type *procurrens* includes tall plants with large inflorescence. Pinnate leaves with long petioles and few, relatively wide leaflets (end leaflets broader than side leaflets) that are strongly indented. Large flowers and hairless fruit. Runners are formed above and below ground.
4. Basic type *sambucifolia* is small with small inflorescences. Pinnate leaves with relatively short petioles and a small number of leaflets which are wide, stemmed, with many broad teeth. No hairs on stems or fruits, flowers and fruit fairly large. Plants form runners above and below the surface.

Taken from: Jochen Bockemühl, *Ein Leitfaden zur Heilpflanzenerkenntnis*, Dornach: Verlag am Goetheanum, II, Dornach: Verlag am Goetheanum, 2000 (English translation in preparation).

Source of illustrations

Drawings and colour illustrations are by Jochen Bockemühl apart from: flower development, pages 78, 82, 95 and 98 by Kari Järvinen and common horsetail development, page 105 and 107 by Günter Heuschke.

This book is the result of many years of the authors' collaboration in seminars for farmers on Rudolf Steiner's Agriculture Course.

Biodynamic Agriculture

Willy Schilthuis

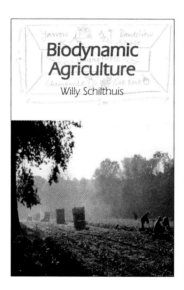

A concise and fully illustrated introduction to biodynamics. Biodynamics is an internationally recognized approach to organic agriculture in which the farmer or gardener respects and works with the spiritual dimension of the earth's environment.

In a world where conventional agricultural methods threaten the environment, biodynamic farms and gardens are designed to have a sound ecological balance. This book presents evidence that biodynamic crops put down deeper roots, show strong resistance to disease and have better keeping qualities than conventionally produced crops.

www.florisbooks.co.uk

Principles of Biodynamic
Spray and Compost Preparations

Manfred Klett

A renowned biodynamic expert, Klett provides a fascinating overview of the history of agriculture, then goes on to the discuss the practicalities of spray and compost preparations and the philosophy behind them.

This is essential reading for any biodynamic gardener or farmer who wants to understand the background to core biodynamic techniques.

Based on keynote talks given by Manfred Klett at Biodynamic Agricultural Association conferences.

www.florisbooks.co.uk